The Political Economy of Environmental Protection

The Political Economy of Environmental Protection

Analysis and Evidence

Roger D. Congleton, Editor

Ann Arbor

THE UNIVERSITY OF MICHIGAN PRESS

Copyright © by the University of Michigan 1996
All rights reserved
Published in the United States of America by
The University of Michigan Press
Manufactured in the United States of America
⊚ Printed on acid-free paper

1999   1998   1997   1996   4   3   2   1

A CIP catalog record for this book is available from the British Library.

Library of Congress Cataloging-in-Publication Data

The political economy of environmental protection : analysis and
    evidence / Roger D. Congleton, editor.
        p.   cm.
    ISBN 0-472-10602-3 (hardcover : alk. paper)
    1. Environmental policy—Economic aspects.   I. Congleton, Roger D.
HC79.E5P656   1996
333.7—dc20   96-25
                                                              CIP

This book is dedicated to the memory of my wife Caroline,
a wonderful woman and constant friend
who left flowers in her footsteps

Preface: The Political Economy of Environmental Protection—Analysis and Evidence

Environmental analysts agree that many of the by-products of modern production impose health and esthetic costs on people in the areas surrounding effluent sources. There is much less consensus about the magnitude of the damage or the extent of the regions affected. Environmental analysts also agree that markets tend to produce Pareto excessive quantities of undesired effluents. There is much less consensus about the speed and extent to which effluent reducing policies should be adopted, on the steps that should be taken, and on the effectiveness of policies already adopted. These areas of controversy are a partial explanation of the controversial nature of environmental policies among experts and voters.

A lack of consensus about environmental policies by the well informed implies that political processes by which conflicting opinions are weighed ultimately determine policy. Different decision rules and/or procedures will yield different environmental laws. These processes must be examined if current and future policies are to be understood. Moreover, it may be argued that *political institutions fundamentally determine environmental quality* inasmuch as environmental laws affect rates of return for alternative production technologies and thereby patterns of investment and the general level of economic activity.

The articles included in this volume analyze, and in many cases empirically verify, the pressures within and between governments that affect environmental policies. Generally the empirical results indicate that the policy-setting framework is neither fully captured by special interest groups nor fully determined by electoral pressures. Environmental policies and thereby environmental outcomes emerge from a complex process of political accommodation of conflicting interests. Consequently, observed environmental policies do not maximize social net benefits nor do they simply implement the median or average voter's environmental preferences.

My aim in assembling this volume is to reflect the broad mainstream of modern political analysis of environmental policy. Each contributor has developed a balanced and generally positive treatment of the political and economic issues addressed. The authors are unanimous in their assessment that both political and economic considerations affect environmental policies. Many of the chapters are necessarily technical, but the general thrust of these analyses can be readily discerned without carefully reviewing the authors' scientific methodology. The work included should be of interest to the broad spectrum of academics interested in the politics of environmental regulation.

The volume includes four previously published articles from leading economics journals and eight previously unpublished works by economists and political scientists. Although the cases examined are largely from the United States, the methods of analysis are relevant for other regions of the world, once properly modified to account for relevant variations in local institutions. A wide range of modern research methodologies is employed to analyze several related political issues. What kinds of environmental policy remedies are most likely to be chosen (chaps. 2 and 3)? How persuasive are the efforts of special interest groups? Are campaign resources and the votes of special interest groups so crucial to a candidate's electoral success that candidates must promote the interests of well-organized interest groups if they are to be reelected (chap. 4)? Are elected representatives and the bureaucracy so constrained that they always adopt policies that maximize the median or average voter's welfare? Or is electoral competition sufficiently weak, and environmental issues of little enough salience, that representatives and the bureaucracy are free to maximize their own environmental agendas without regard to the interests of voters (chaps. 5, 6, and 7)? How do anticipated environmental policies affect the decisions of firms (chap. 9)? How do governments deal with externalities that extend beyond the boundaries of their jurisdictions (chaps. 8, 10, 11, and 12)? Only two of the chapters directly address the normative issue of whether the resulting policies are efficient or effective. A more complete examination of these normative issues is of interest but is left for future work and another collection.

It is important to remember that the interaction between political and economic spheres is not unidirectional. Economic growth and development generate many of the environmental problems to be addressed but they also generate the means for dealing with these problems and often the political impetus for policy decisions. The latter is most obvious in the case of economic interest groups who oppose stringent environmental rules on the basis of anticipated compliance costs. The resources at the disposal of politically active environmental interest groups are also partly a matter of the economic wherewithal of their subscribers and sponsors. Current environmental policies, and those likely to be adopted in the future, are best be understood as a joint consequences of political and economic factors.

The chapters in this book provide analytical and empirical support for this contention. The policies examined by the contributors have only been indirectly affected by notions of economic efficiency. Rather, the diverse perceptions of voters and conflicting goals of politically active groups determine policies by their participation in the political system. Economic and political interests are found to affect the environmental policies adopted by local and national governments, and those negotiated in international treaties between national governments.

Worldwide interest in prospects for human-induced climate change suggests that environmental matters will remain a significant area of regulatory debate for the foreseeable future. The political economy of pollution control is similarly bound to remain on the research agenda for years to come as we attempt to better understand the processes by which environmental policies are adopted and implemented.

Acknowledgments

The International Institute of George Mason University has provided long-term support for my work on environmental politics, including support for this volume. The help and encouragement of John Moore, James Buchanan, and Caroline Spalding have made this volume possible. Part of the task of assembling materials and writing introductory materials was done while a visiting fellow at the Research School for Social Science at Australian National University. Thanks are due to the contributors and to many others at GMU and ANU for stimulating conversations and helpful suggestions. Carol Robert did yeoman work formatting and organizing final corrections.

Contents

Political decisions largely determine man's relation to the environment
by specifying the scope of lawful means by which natural resources
may be employed. Environmental laws may exacerbate or moderate the
relationship between pollution and economic growth according to the
rules chosen. As is the case for other types of legislation, the politics of
environmentally relevant laws reflects trade-offs between the diverse interests of individual voters and the pressures of economic and environmental interest groups. The papers in this book examine the politics of
environmental law.

Buchanan and Tullock develop a positive theory of externality control
which explains the relative frequency of direct regulation as opposed to
penalty taxes or effluent charges. They argue that in most cases industrial groups will prefer standards to effluent charges because the former
tend to generate economic profits for the affected firms while the latter
do not.

Leidy and Hoekman explore the extent to which international trade concerns may be expected to affect domestic and international environmental
policies. They argue that inefficient environmental regimes strengthen
the trade-policy linkage in several ways. Inefficient environmental policies

set a precedent for market sharing and establish barriers to entry that reduce the dissipation of profits associated with protectionist trade policies. Consequently, firms in either a large or small open economy have even stronger reasons to lobby for inefficient environmental policy measures and output controls than those identified by Buchanan and Tullock.

nomic costs and benefits. However, their results reveal that the testimony of interest groups also affected the probability that a particular pesticide use could be continued. (*Journal of Political Economy* 100: 175–97)

Hoagland and Farrow examine the extent to which economic analysis and political interests have affected Department of Interior decisions regarding the sale of offshore drilling leases. The offshore leasing process takes place in two stages. In the first stage, tracts of undersea land are evaluated for their marine and geological properties, hearings are held, and a subset of the range of possible lease sites is announced. Subsequently, the original list of sites may be modified as leases between the federal government and oil firms are actually consummated. Their analysis suggests that interest groups' activities played an important role in both stages of the leasing process but had stronger effects on the leases actually consummated.

Part 4. Relations between Governments

Faulhaber and Ingberman analyze properties of alternative institutional solutions to a local government NIMBY game. They examine three institutional arrangements for locating waste facilities: (i) market, (ii) centralized democratic government, (iii) centralized democratic government with host veto. The assignment of bargaining power turns out to be crucial to the efficiency of these alternative regimes. If information is observable and the potential host community has bargaining power, then either competitive or monopoly markets for the NIMBY activity (waste disposal) generate efficient location of facilities. Similar results are obtained for majoritarian central government.

Hamilton explores the extent to which political and economic circumstances have affected the decisions by waste disposal companies to locate their facilities. He argues that waste disposal firms can lower their overall costs by locating in areas where political resistance is likely to be smallest. Estimates of locational choice functions reveal that, other things being equal, companies tended to choose areas where environmental interests are least likely to be effectively promoted by local citizen groups. (*Rand Journal of Economics* 24:126–46)

Part 5. Environmental Treaties

Contributors

Professor Stephen J. Berardi
U.S. Department of Treasury
Washington, D.C.

Professor James M. Buchanan
Center for Study of Public Choice
George Mason University

Professor Dennis Coates
Department of Economics
University of North Carolina

Professor Roger D. Congleton
Center for Study of Public Choice
George Mason University

Dr. Maureen L. Cropper
World Bank
Washington, D.C.

Professor Maria M. Ducla-Soares
Department of Economics
National University of Lisbon

Professor William N. Evans
Department of Economics
University of Maryland

Professor Scott Farrow
H.J. Heinz II School of Public Policy
Carnegie Mellon University

Professor Gerald R. Faulhaber
Wharton School
University of Pennsylvania

Professor Lars P. Feld
University of Saarland
Saarbrüeken, Germany

Professor James T. Hamilton
Terry Sanford Institute of Public Policy
Duke University

Professor Albert Hart
University of Saarland
Saarbrüeken, Germany

Professor Porter Hoagland
Graduate College of Marine
 Studies
University of Delaware

Dr. Bernard M. Hoekman
World Bank
Washington, D.C.

Professor Daniel E. Ingberman
John M. Olin School of
 Business
Washington University

Dr. Michael P. Leidy
International Monetary Fund
Washington, D.C.

Professor Werner W.
 Pommerehne
University of Saarland
Saarbrüeken, Germany

Dr. Paul R. Portney
Resources for the Future
Washington, D.C.

Professor Todd Sandler
Department of Economics
Iowa State University

Professor Gordon Tullock
Department of Economics
University of Arizona

Professor George Van Houtven
Department of Economics
East Carolina University

Part 1
Introduction

CHAPTER 1

Introduction and Overview of the Political Economy of Environmental Protection

Roger D. Congleton

We are in a position more and more completely to say how far the waste and destruction of natural resources are to be allowed to go on and where they are to stop. It is curious that the effort to stop waste, like the effort to stop forest fires has often been considered as a matter controlled wholly by economic law. I think there could be no greater mistake.
—Gifford Pinchot (*The Fight for Conservation* 1910)

Introduction: Conceptual Foundations of Environmental Politics

Environmental policy concerns all matters which directly or indirectly affect the manner and the extent to which human society makes use of nonhuman resources. Rules which allow or promote hunting and farming imply that some fauna and flora become more common and others less so. Property rights which encourage mining and manufacturing imply that some minerals become less common and others more so. Mankind cannot sustain itself without relying on nature for fundamental necessities, and any use of natural resources, naturally, alters the environmental balance in discernible ways.

The role of politics in determining the manners in which natural resources may be utilized is pervasive. Political decisions largely determine the feasible uses of natural resources by defining and enforcing fundamental property rights and entitlements over matters with environmental consequences. Government policies affect population magnitudes and densities through effects on birth rates, mortality rates, land use, and patterns of migration. Government policies affect the environment through policies and expenditures which affect the rate of return and thereby the accumulation of new knowledge and technologies. By determining the rules of the game, political deci-

3

sions determine the extent to which humanity is empowered to transform the world, the processes used, and the waste products produced.[1]

Only a subset of the broad range of environmentally relevant policies is principally motivated by environmental concerns. These policy areas are often referred to as environmental law or environmental protection. The chapters in this volume focus on the political and economic underpinnings of such policies. Other recent volumes have addressed economic and game theoretic aspects of environmental policies, but have neglected the political processes by which environmental policies are actually adopted and implemented. This book remedies that neglect by examining political factors which influence environmental policies. Analysis of the politics of broader policies which have environmental ramifications is, of course, also of interest, but focusing on environmental policies per se reduces the number of factors to be examined to manageable numbers. This allows a more systematic analysis and facilitates empirical testing of the analyses developed.

The first section of this introductory chapter addresses several conceptual issues raised by environmentalists in the greater green debate who approach environmental policy as a "crisis area" rather than as an ongoing policy concern. The second section develops a rational choice perspective on environmental politics in order to provide a frame of reference for the papers included in the volume. The third section develops an overview and summary of the papers included in the volume. The last section provides a few concluding remarks.

Environmental Controversy

The focus of the various policies of "environmental protection" is not really about protecting the world's nonhuman resources from mankind, as the term seems to suggest. Nor is environmental protection an attempt to minimize the effect of human activities on nonhuman natural processes.[2] While it may be technologically feasible to completely isolate mankind from nature in a hermetically sealed self-sufficient enclave, all but the most extreme environmentalists envisage a world in which man continues to flourish and range over a largely unrestricted domain. That is to say, environmental policies take for granted the existence of some greater than minimal human impact on the environment. At root then, environmental policy is a matter of determining the "appropriate use" of natural resources by mankind.

Most environmental controversies concern alternative interpretations of the phrase "appropriate use," and the political processes that determine which

---

1. See Dryzek 1987 for a thoughtful and ambitious green perspective on the scope of a comprehensive environmental policy.

2. Some environmentalists seem to make this argument. See, for example, Goodin 1992.

interpretation(s) will be implemented as policy. Opinions differ widely about the "appropriate use" of natural resources for a variety of reasons. Individuals disagree about the role of nature in what might be called "the good life." Some individuals are highly appreciative of nonhuman processes of transformation. Others find nature disorderly, dirty, and dangerous. Clearly the untrammeled original "natural" state is more attractive to those who believe that material comforts impoverish the soul, than for those who believe that physical discomfort distracts one from higher ends. Disagreements also exist over assessments of man's responsibility to other species, and over such scientific matters as the resilience and stability of meteorological and ecological systems.

Differences in economic circumstances and comparative advantage also play a role in determining the range of opinion. To a builder, the "appropriate use" of a piece of land may be to provide a comfortable habitat for humans through appropriate modifications in the location, arrangement, and assortment of natural and man-made resources present at the site and elsewhere. A builder might wish to clear the "site" and build houses. To a farmer, its "appropriate use" may involve a somewhat less radical transformation of the existing state of nature, although it may also entail removing current residents from the site and moving in new residents—clearing, plowing, and planting corn. To a naturalist, its "appropriate use" may entail erecting legal or physical barriers around the site to exclude those who have other ideas about "appropriate use."

Stated in an economist's terms, *individuals disagree about the extent to which various human and nonhuman processes of transformation add or decrease value*. These differences lead individuals to adopt different levels of "environmental protection" in their own lives—making more or less use of pesticides, motor cars, sidewalks, imported flowers, and the like—and to disagree about the relative merits of public policies which affect the manner and extent to which natural resources may be utilized. It is these differences which political arrangements confront when making policy decisions. And it is these disagreements which make the political economy of environmental protection a subject of interest to those who wish to understand the political and economic processes which characterize man's lawful relation to nature.

The Problem of Environmental Management Is Not New

Although the volume of environmental regulation in the United States has increased substantially during the past thirty years, these laws are by no means the first American efforts to adopt rules for the purpose of restricting or redirecting man's use of the environment. As early as 1626, the Plymouth Colony passed ordinances regulating the cutting and sale of timber on colony lands (Meyer 1966). In 1652, the first public water supply was constructed in Boston. In 1657, the burgomasters of New Amsterdam issued an ordinance prescribing that the streets be kept clean and that all rubbish and filth be deposited at certain designated places (Sopper 1966). In 1681, William Penn required that new landowners leave an acre of forest standing for every five acres cleared in his ordinance for the disposal of lands (Meyer 1966).

Rules to control access to common property resources, to regulate waste disposal, and to assure potable supplies of water, of course, predate these American efforts by many centuries. Appreciation of the environment's role in health and beauty coincide with the dawn of written history. To take a relatively modern example, well after sunrise, consider Aristotle's discussion, in passing, of policies concerning water and air quality in his characterization of the ideal community.

> I mention situation and water supply in particular because air and water, being just those things that we make most frequent and constant use of, have the greatest effect on our bodily condition. Hence in a state which has [the] welfare [of its citizens] at heart, water for human consumption should be separated from water for all other purposes. (1969, p. 278)

Community waste disposal sites and burial grounds have greatly facilitated the research of archeologists in what surely must be an unanticipated consequence of the environmental regulations adopted by the civilizations of antiquity.

Environmental prerequisites to a comfortable and healthy life have long been a practical matter at the core of economic prosperity.[3] What has varied through time has been the efficacy of efforts to mold nature into more useful and pleasing forms as the aesthetic assessments and technological feasibility of alternative uses of natural and human resources have varied. The fundamental environmental policy issues are today, as ever, where to draw the line and where the line will be drawn.

A Digression on Technological and Legal Foundations of the Environmental Commons

Natural resource management issues may usefully be subdivided into two areas: those regarding the management of stationary resources and those regarding the management of freely circulating resources. Both these areas

---

3. Natural phenomena play such an important role in both hunting and agricultural based societies that very often nature has been used as the basis of extensions into metaphysical and religious realms. Even today, various forms of nature worship or pantheism are among the most common world religions.

have long been subjects of government regulation, and both exhibit similar problems insofar as free access to either sort of natural resources tends toward excessive use: what Hardin (1968) termed the tragedy of the commons. However, the cost, effectiveness, and feasibility of alternative management methods differ substantially across these two types of resources. Regulating access to stationary resources is generally less problematic than regulating access to freely circulating resources because monitoring costs are lower and the problems associated with mismanagement are more obvious.

Historically, a variety of management methods have been used to limit access to stationary resources in order to promote their productive and sustained use. One broad class of methods widely used for centuries involves the assignment of "use rights" for particular stationary resources, such as grazing lands, gardens, or lodging sites. Here the political authorities do not directly regulate the use of natural resources, but rather define and protect the use rights of lawful "owners."[4]

In cases where "use rights" are exclusive *and* tradable, markets for "use rights" give owners incentives to consider both the current and future productivity of the resource over which they have control. The current income and resale value of a well-managed site tends to be higher than one that is poorly managed. Exclusive and marketable "use rights" also create a low-cost method whereby resources may be shifted from more talented resource managers. Better managers are naturally willing to pay a higher price for use rights than poor managers because they anticipate greater output from the same resources.

Few would argue that privatization solves all the problems of managing stationary resources, but the efficiency of the property rights solution as a method of encouraging the production of valued services for humankind is attested to by the fact that formal chains of title to real estate in Western Europe and in the Eastern United States generally predate current governments, often by several centuries.[5] The property rights solution to the management of stationary resources is relatively permanent and unobtrusive and

---

4. Ostrom (1990, chap. 3) analyzes a variety of long-standing methods of managing access to common property resources, including grazing lands in the Swiss Alps, communal forest land in Japan, and canal systems in Spain. She concludes (p. 90) that enduring methods for resolving commons problems share eight characteristics: (1) clearly defined boundaries, (2) congruence between appropriation and provision rules and local conditions, (3) collective choice arrangements, (4) monitoring for appropriate behavior, (5) graduated sanctions, (6) conflict resolution methods, (7) minimal rights to organize locally, and for larger commons problems, (8) organization of monitoring, enforcement, and conflict resolution as "nested" enterprises.

5. See Macfarlane 1978 for an overview of English property law during the medieval period.

requires little political oversight on a day-to-day basis.[6] Monitoring the unauthorized use of privately owned resources is largely undertaken by property owners themselves, which further reduces the cost of this solution to the commons problem.

Management of freely circulating resources is more problematic and is, partly for this reason, of greater modern policy interest. Circulation and diffusion give air and water systems a substantial capacity for dissipating and neutralizing many of the undesired byproducts of farming, manufacturing, and transport. Circulation and diffusion also imply that access to these resources is largely unimpeded by inconvenient distances, no trespassing signs, or fences. These properties have long been relied on (at least implicitly) as inputs in the production process. It is, after all, what makes fire, farming, and indeed breathing, possible. Unregulated free access to these systems eventually leads to overexploitation of their capacities in much the same manner as was true for stationary resources. New users gain average net benefits from use rather than marginal net benefits. Marginal net benefits are below average benefits (and may even be negative) because a large proportion of productivity losses are borne by other users in the form of reduced air or water quality.

The same properties of the air and water systems which make them valuable as inputs for production and as a vehicle for waste disposal also make them difficult to regulate and/or to privatize. Communities cannot assign use rights to *particular units* of freely circulating air and water in the same manner that they can for land and mineral resources.[7] Air and water are unlike railroad boxcars and cattle in that particular "pieces" of air or water cannot be readily identified, isolated, or therefore assigned to particular "owners." This contrasts with, for example, homogeneous acreage in a featureless desert or plain which can be assigned to owner/users on the basis of geometric coordinates. Tradable rights to *use* air and water can be established, but such rights are *user*

---

6. The above argument is not meant to imply that privatization has very often been adopted with the conscious intent of solving commons problems. Rather, it is likely that regions of the world which adopted ownership rights avoided commons problems that others societies using more politicized methods had to cope with. Through time, as legal institutions evolved, better assignments of use rights tend to supplant management methods which yielded significantly inferior outputs.

It may be argued that there has been a gradual restriction of private use rights through time. Often these restrictions do not substantially restrict the uses of the stationary resource, but rather access to *circulating* resources which pass under, over, or through the stationary resource. For example, modern rules governing solid waste disposal attempt to ensure that "disposal" is undertaken in a manner which does not affect local water or air quality.

7. One can own bottles of air or water. But, generally, bottled air and water are only temporarily partitioned from the common circulating air and water systems. "Purchase" of a bottle of water is more analogous to renting than owning an asset.

rights to a common property resource rather than ownership in the usual sense.[8]

Politics in areas of communal resource management is nearly unavoidable. Controversy over appropriate access to communal resources is likely because common ownership implies that decisions to curtail, maintain, and/or increase "use rights" directly affect the interests of all other users of the common property resource. The physical properties of circulating natural resources imply that a relatively large number of affected parties will disagree over the importance of controlling particular point sources of effluent emissions. Diffusion generally diminishes the impact of effluents as distance from the point of origin increases. Consequently, even if there were no other disagreement regarding acceptable emission rates, diffusion provides a physical basis for political controversy at all levels of government with authority over the affected parties. (Faulhaber and Ingberman analyze some political implications of such effects in chap. 8 of this volume.)

The Politics of Environmental Protection

The aim of this section of the introduction is to develop a frame of reference for the more particular analyses of the rest of this volume. All the papers in this volume use the tools of modern political science and economics to analyze various determinants of specific policy areas. None of the authors explicitly address all the complex relationships between the politics and economics of environmental regulation, but rather develop as narrow or abstract a model as is useful for the subject examined. The framework set forth here provides an overarching model of political/economic interaction which allows the papers to be linked up to form a general analysis of environmental politics.

Environmental policies reflect the interests of those empowered to make policy decisions given their personal, economic, and political constraints. Constitutional and legal arrangements determine which individuals are empowered to make environmental policy decisions in a democracy and to whom such policy makers are accountable. Economic considerations determine the real cost and feasibility of alternative policies. These political and economic constraints may be so binding that policymakers have little discretion over policy. Elected representatives and the bureaucracy may unfailingly advance the perceived policy interests of the electorate. Or, political and economic

---

8. See Block 1990 for several extended discussions of the merits of market-based environmental regulation. These discussions generally neglect the fact that determining the optimal quantity of use permits, and/or the range of uses (permissible effluent rates) allowed are bound to remain ongoing political/regulatory issues.

constraints may allow so much discretion that government decision makers simply do as they please on environmental matters.[9]

Inasmuch as the environmental policies of interest are generated by human actions, all politically relevant demands for environmental quality are at root personal demands. Consequently, models of individual choice in various political and nonpolitical settings can be used to analyze the determinants of environmental policy. The rational choice approach to political economy is used throughout this volume and is illustrated below with an analysis of an individual's private and political demand for environmental quality. Mathematics is used to develop rigorously an internally consistent model of economic and political relationships but is not essential to the perspective developed.

A Model of Environmental Choice: The Private Demand for Environmental Quality

Suppose that an individual must allocate some personal holding of natural resources (perhaps a piece of forest land) between its current natural state valued for its own sake and a production process which yields desired outputs and undesirable effluents. The environmental tradeoff is represented with a production function that includes effluents as an "input" into the production of desired outputs, $Y_i = f(R_i^p, E_i, T)$. Desired output, Y_i, increases as more natural resources are devoted to production, R_i^p ($Yi_{R_{pi}} > 0$). And, for any given use of natural resources, desired output increases as more effluents, E_i, are released ($Yi_{E_i} > 0$).[10] Advances in technological knowledge, T, increase the extent to which desired outputs can be obtained from a given use of natural resources ($Yi_{TR_{pi}} > 0$) and diminish the extent of effluents that need be generated ($Yi_{TE_i} < 0$). The resources that the individual devotes to production (R_i^p) reduces what remains in its natural state for other purposes ($R_i^n = R_i^o - R_i^p$).[11]

---

9. The role of political institutions in determining discretion can easily be made concrete. Imagine the different circumstances of a dictator and member of the House of Representatives. The level of support that the representative requires to be maintained in office is clearly greater than that required for a dictator. Consequently, dictators have greater discretion over environmental policies than elected representatives have.

10. Such a tradeoff is typical of many primitive and sophisticated production methods. A farmer may get more farm crops using "slash and burn" than by carefully harvesting the forest first, for a given use of his resources, but "slash and burn" generates more effluents than carefully harvesting the forest would have.

11. Natural resources are assumed to have the properties of a pure private good and be privately held for the purposes of this exercise. Any notional value placed on the total stock of natural resources is neglected, as are various negative externalities associated with maintaining resources in their natural state (pests and predators). Moreover, other external effects of using

The individual's allocative decision is represented as an attempt to maximize a utility (objective) function defined over natural resources (possibly including leisure), produced goods, and effluents, given existing production technology, T, environmental knowledge, I, and a personal endowment of natural resources, R_i^o. Individuals control the extent to which their own resources are used to produce desired and undesired outputs (Y_i and E_i), but not the overall level of effluents confronted by the individual. The ambient pollution level, E, is the sum of all producers in the area of interest, $E = \Sigma_{i=1}^N (E_i)$. The repulsiveness of this pollution varies with its level and the individual's knowledge, I, of its harmful or unattractive nature.

In order to simplify exposition, without significant loss of generality for the purposes of this illustration, each decision variable is treated as if it is single dimensioned, although it is clear in fact that each is multidimensional. The effects of time and uncertainty are also neglected, or subsumed in the functional form of U_i.[12] As in the above, the resource use and output of a typical individual is characterized as an effort to maximize

$$U_i = u_i(R_i^n, Y_i, E, I), \tag{1}$$

subject to

$$Y_i = f(R_i^p, E_i, T), \tag{1.1}$$

$$E = \sum_{i=1}^N E_i, \tag{1.2}$$

and

$$R_i^o = R_i^n + R_i^p. \tag{1.3}$$

Substitution allows the decision to be cast in environmental terms as a choice of nonproduction uses of resources and effluent emissions.

---

natural resources in manufacture are neglected. Harvesting a field or a forest may leave an unattractive stubble. These non-effluent types of externalities are neglected here in order to focus attention on decisions affecting effluent emission and subsequent diffusion.

12. Uncertainty and time can be incorporated in the model explicitly by characterizing an intertemporal and stochastic environmental trade off. For example, rather than $Y_i = f(R_i^p, E_i, T)$, one could use $E_i \sim f(Y_i, R_i^p, T, t)$. In this case, environmental damages are uncertain; they depend on time, t, production of the desired output, Y_i, technology, T, and production use of the natural resource, R_i^p. Expected utility is $U^e = \int\int u_i(R_i^n, Y_i, E, I, t)f(Y_i, R_i^p, T, t)\,dt\,dE$. Note that this implies that, for a given technology, expected utility can be written as a function of R_i^n, Y_i, E, and I.

$$U_i = u_i \left(R_i^n, f(R_i^o - R_i^n, E_i, T), \sum E_i, I \right) \qquad (2)$$

Differentiating with respect to R_i^n and E_i and setting the result equal to zero allows the individual's utility-maximizing private environmental policy to be characterized as that combination of R_i^n and E_i which satisfies

$$U_{Rn} = U_Y Y_{Rp}, \qquad (3.1)$$

$$-U_E = U_Y Y_E, \qquad (3.2)$$

or, dividing,

$$U_{Rn}/-U_E = Y_R/Y_E. \qquad (3.3)$$

The individual's personal choice of environmental quality is determined jointly by objective and subjective tradeoffs. Generally, the greater are the objective effects of alternative effluent and natural resource rates on personal income, the less interest individuals have in preserving natural resources or limiting their emissions of effluents. The greater the marginal appreciation of natural resources for their own sakes or marginal distaste for effluents, the less inclined individuals will be to devote natural resources to manufactured goods.

Equation (3.3) summarizes these tradeoffs. Resources are retained in their natural state (withheld from production) and effluents are emitted in the combination where the marginal rate of (private) substitution between natural resources and pollution equals their technological rate of substitution in production. The implicit function theorem allows the individual's preferred use of natural resources and effluent emissions (pollution) to be characterized as functions of the various parameters of the individual's optimization problem:

$$R_i^n = r_i \left(R_i^o, \sum_{j \neq i}^{N} E_j, T, I \right) \qquad (4.1)$$

and

$$E_i = e_i \left(R_i^o, \sum_{j \neq i}^{N} E_j, T, I \right). \qquad (4.2)$$

Together, these equations characterize the individual's private demand for environmental quality as functions of various parameters of the individual's choice problem. Desired environmental quality varies with the extent of natural resources controlled, technological and cultural knowledge, and the effluent emissions of other individuals.[13]

These "best reply" functions can be used to characterize overall equilibrium levels of private environmental quality. A private environmental quality (Nash) equilibrium occurs when all individuals in the region of environmental interdependence have simultaneously chosen their utility-maximizing use of effluent emissions and natural resource usage, given those of other relevant individuals.

A good deal of normative environmental analysis stresses the fact that the likely Nash equilibrium of the above private environmental quality game is not Pareto optimal. Consequently, mutual gains to trade exist which might be realized through government policies which coordinate or control individual propensities to pollute. (Werner Pommerehne in chap. 10 of this volume explores an unusual case where these mutual gains are realized through a contracting process.) On the other hand, previous work has largely neglected the fact that *political interest in environmental protection would exist even in the unlikely case that the private environmental quality equilibrium is Pareto optimal.*

The Political Demand for Environmental Protection

An individual's political demand for environmental quality differs from his private demand in that one's political demand for environmental quality does *not* have to take the behavior of other polluters as given. At the political level, the environmental choices of other individuals can be controlled, or at least influenced, by the coercive powers of government. In other details, assessment of an individual's preferred environmental policy is fundamentally similar to the individual's private choice characterized above. Environmental policy preferences are again the outcome of a combination of objective and subjective effects of policy alternatives.

Some insight into an individual's policy predilections can be obtained by examining the tradeoffs faced by a typical individual in choosing parameters of a specific environmental policy. Suppose that an individual is attempting to determine his most preferred combination of "Pigovian" taxes on effluent emissions and natural resource use. Taxes are to be imposed on all natural resources used in production (t^p), and/or on all effluent emissions (t^e). As generally assumed in the public finance literature, the tax receipts are to be returned as demigrants to all taxpayers within the government's jurisdiction

---

13. Differentiating with respect to R_i^o, T, and I allows the comparative statics of the individual's demand for environmental quality to be characterized.

$(G = [\Sigma t^p R_i^p + \Sigma t^e E_i]/N)$. One effect of the environmental taxes is to increase the marginal cost of manufactured goods at the individual level. Another is to increase the wealth of individuals who use below average natural resources in production and emit below average levels of effluents. Other uses of the tax revenue would have implied similar, if more subtle and indirect, distributional effects. Both the environmental and redistributive effects will influence policy preferences.

The individual's manufactured income constraint, equation (1.1), becomes $Y_i = f(R_i^p, E_i) - t^p R_i^p - t^e E_i + G$. Pigovian user taxes reduce the rate at which desired outputs can be obtained from inputs, which tends to decrease the extent to which resources are used in production (ΣR_i^n rises), and decreases effluent emissions (ΣE_i falls). The taxes simultaneously affect the choices of all individuals in the polity, and thereby the Nash equilibrium resource use and effluent emissions. Consider an individual at a pretax equilibrium,

$$U^* = u\Big(R^{n*}, f(R^{o*} - R^{n*}, E^*, T) - t^p(R^{o*} - R^{n*}) - t^e E^*$$
$$+ \Big[\sum t^p(R^{o*} - R^{n*}) + \sum t^e E_i\Big]\Big/ N, \sum E_j^*, I\Big), \tag{5}$$

where the "starred" variables denote demand functions as developed above for effluent emission and natural resource, augmented to include tax arguments.[14] Differentiating with respect to the two taxes, and appealing to the envelope theorem, yields first order conditions describing the individual's preferred "Pigovian" tax scheme:

$$U_Y\Big[-(R^{o*} - R^{n*}) + \sum (R^{o*} - R^{n*})/N\Big] + U_E E_{tp}^* = 0 \tag{6.1}$$

---

14. These augmented reaction functions are calculated by replacing the original production relationship in equation (1.1) with $Y_i = f(R_i^p, E_i) - t^p R_i^p - t^e E_i + G$, and adding a new constraint representing the assumed relationship between tax receipts and the demigrant program $[G = (\Sigma t^p R_i^p + \Sigma t^e E_i)/N]$ to the original private optimization problem. The private demand for personal holdings of natural resources and effluent emissions are determined as before to characterize augmented reaction functions of the form

$$R_i^n = r_i\Big(R_i^o, \sum_{j \neq i}^{N} E_j, T, I, G, t^p, t^e\Big)$$

and

$$E_i = e_i\Big(R_i^o, \sum_{j \neq i}^{N} E_j, T, I, G, t^p, t^e\Big).$$

and

$$U_Y \left[-E^* + \sum (E_j^*)/N \right] + U_E E_{te}^* = 0. \tag{6.2}$$

In both equations, the first term characterizes the net marginal cost of the tax in terms of the marginal utility of reduced personal consumption of desired manufactured goods (which can be thought of as personal income). The second characterizes the marginal utility from the overall reduction in emissions caused by the tax.

Note that these first order conditions imply that *even in cases where the environmental effects of "environmental" policy tools are negligible at the margin, the preferred tax rate may, none the less, be greater than zero.* Individuals who receive positive net receipts from the environmental tax and demigrant program as a whole will prefer a relatively more aggressive "environmental" policy for pecuniary reasons. Individuals who lose at the margin from the environmental tax and demigrant program will prefer a relatively less stringent "environmental" policy. Only the average voter, who receives no net transfer from the tax/demigrant policy, evaluates policy parameters strictly in terms of their effect on the perceived environmental quality. The redistributional effects of environmental policy influence voters and their representatives, and it may motivate the activities of politically active interest groups.

Application of the implicit function theorem to equations (6.1) and (6.2) allows an individual's ideal Pigovian tax program to be characterized as a function of parameters of his optimization problem.

$$t_i^{e*} = r_i(R_i^o, T, I, N) \tag{7.1}$$

and

$$E_i^{n*} = e_i(R_i^o, T, I, N). \tag{7.2}$$

As in the case of the private demand for environmental quality, the preferred public policy varies with one's initial natural resource wealth, R_i^o, the technology of production, T, and knowledge of the dangers, I, or lack thereof, of effluent levels. The political demand is also affected by the number of effluent emitters, N. The shape of the demand for environmental fees (taxes) is affected by the shape of the individual's own utility function, those of all others in the relevant polity, and the production function for manufactured goods.[15]

---

15. Differentiating with respect to R_i^o, T, I, and N would again allow the comparative statics of the individual's political demand for environmental quality to be characterized.

As time passes and parameters of the individual's choice problem change, an individual's interest in environmental quality will also change.

Political Institutions, Discretion, and Environmental Politics

In deterministic pure voting models of politics, the median voter's preferred environmental tax policy is the policy adopted.[16] In the above model, if the median voter bears close to the average tax burden, the program will generate the median voter's preferred level of environmental quality. If he bears less than the average tax burden he will tend to prefer higher than optimal tax levels and more stringent than optimal levels of environmental quality.

Richer models of political processes take into account other institutional features of modern democracies which allow individuals to affect political outcomes through such means as lobbying or campaign contributions. Institutional details and the technology of persuasion determine the relative importance of votes cast and dollars spent in such models. If voters are well informed and largely uninfluenced by organized efforts at persuasion, then voter preferences are decisive insofar as representatives are interested in being reelected. In this case, and in the case where interest group pressure is symmetric about the median, median voter preferences will determine policy. If voter opinion is perfectly malleable, then dollars spent during and after the campaign are decisive. See Becker 1983. In the more likely intermediate cases, nonvoting avenues of political pressures may cause policies to diverge from median voter preferences.

Political and legal institutions affect the methods chosen by individuals who endeavor to influence environmental policies by affecting the marginal cost-effectiveness of alternative methods of influence. In democracies, it is generally less costly to cast votes than to testify before relevant legislative committees or regulatory commissions. Consequently, more individuals cast votes than participate in regulatory hearings. A modest interest in changing or

---

16. The most enduring models of electoral equilibrium suggest that voters in the middle of the distribution of voter preferences (the median or average voter) tend to get their policy preferences advanced. In order to win elections in winner-take-all districts, candidates are generally drawn toward relatively moderate positions on all issues, including environmental policies, as a prerequisite to winning the election. See Mueller 1989 for an overview of the modern public choice literature. See Black 1958 for the first modern treatment of the median voter theorem. See Enelow and Hinich 1984 for demonstrations that the Nash equilibrium of a contest between uncertain but vote-maximizing candidates tends to be at the mean of the distribution of voter ideal points. In the case of a symmetric distribution of voter ideal points, the median and average voter are the same.

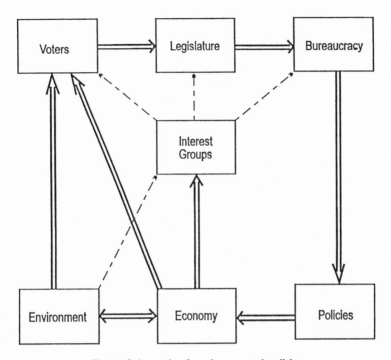

Fig. 1. Schematic of environmental politics

maintaining current environmental policies may be sufficient reason to vote, but not to lobby congress or make contributions to lobbying groups. Those with more intense interests in changing environmental policy will avail themselves of many more channels of influence.

Nonvoting channels of influence exist in democracies because the link between ordinary citizens and environmental policies is indirect. After election to office, representatives may, in principle, vote any way they wish on a piece of legislation. The legislation is subsequently implemented by an unelected bureaucracy.

A single representative cannot be held accountable for the stringency of environmental law because he/she is only one of many charged with developing environmental rules. Moreover, a few unanticipated votes over environmental policies are unlikely to substantially change a representative's future electoral prospects. Individuals vote for representatives based on their anticipated voting behavior over a large number of issues. Even in areas where candidate policy positions are well known, most voters can do little more than choose the least objectionable candidate from among those with a chance of winning the election.[17]

Given the potential for candidate discretion, it is somewhat surprising to find that the median voter model does a good job of predicting many government policy parameters. A substantial body of evidence suggests that the policies adopted by democratic governments are broadly consistent with maximizing the welfare of the median (or average) voter.[18] On the other hand, the evidence is not so great as to rule out at least a modest exercise of representative discretion.[19] In chapter 4, Dennis Coates develops statistical evidence that the campaign contributions of environmental and anti-environmental interest groups, as well as the personal characteristics of representatives, have had modest effects on the voting behavior of elected representatives.

Once environmental legislation is passed, its implementation, and a good deal of the regulatory detail, are left to a largely invisible chain of command within departments and agencies with authority over the matters of interest. Again, at least superficially there appears to be scope for considerable bureaucratic discretion and thus a role for effective interest group activities. There are a variety of reasons why the legislature may grant some discretion to the bureaucracy: the advantages of specialization, the desirability of making use of case-specific knowledge, or congressional desires to shift controversial decisions to another branch of government. Both specific and incidental authority have impacts on environmental legislation insofar as final regulations and guidelines are often drafted by regulatory agencies rather than the legislature. To the extent that bureau decision makers have personal or institutional policy preferences, environmental polices will tend to reflect the economic, aesthetic, and ideological interests of those bureaucratic decision makers.[20]

---

17. This is of course an implication of both the median voter and stochastic/average voter models. The range of opinion on environmental and other issues suggests that only district median voters will see their representative's votes as truly "representative."

18. The median voter model has been widely used in the empirical and theoretical public finance literature to characterize the fiscal policies of democratic federal, state, and local governments. See, for example, Holcombe 1980; Denzau and Grier 1984; and Congleton and Shughart 1990 for empirical evidence on the predictive performance of median voter models of policy formation. None of these models fit the data so well as to rule out other factors. However, the results do suggest that policies do "on average" advance the interests of the median voter. The unexplained residual may be interpreted as evidence of discretionary behavior.

19. A fairly large literature on what has come to be called "Congressional shirking" has been able to find limited (and somewhat controversial) evidence of statistically significant shirking. See Kalt and Zupan 1984. Although congressmen appear to have substantial opportunity for discretionary behavior, they do not appear to ignore the interests of their constituents or to exploit systematically the discretion that they would appear to have.

20. Indeed, individuals especially interested in environmental policy matters may seek out jobs in the relevant bureaus with the express goal of having an impact on policy.

On the other hand, there is a body of theoretical and empirical work that suggests that the various bureaucracies are responsive to the desires of Congress and congressional oversight committees, and thereby to the interest of average or median voters. Environmental policies adopted by the bureaucracy may also, indirectly, reflect electoral considerations.[21] However, the evidence is not so compelling as to rule out all possibility of discretionary behavior by bureaucratic policy makers. Chapters 5, 6, and 7 of this volume suggest that the Environmental Protection Agency (EPA) does exercise some of its potential discretion. Chapter 5 provides evidence of the exercise of unauthorized discretion.

Interest Groups, Persuasion, and Environmental Politics

One widely used method by which interest groups attempt to influence environmental policy is the strategic dissemination of policy-relevant information. Interest groups subsidize the dissemination of information to voters, Congress, and the bureaucracy that highlights the relative merits of their preferred policies.[22] Such information can increase or decrease an individual's demand for environmental quality because definitive information about the merits of alternative environmental policies does not exist. If the information provided is impartial and unbiased, or perfectly filtered by recipients, it would tend to improve the legislative process by reducing voting and regulatory mistakes.[23] On the other hand, to the extent that interest groups are able to manipulate voters and/or their representatives by strategically providing biased or incomplete information, they may be able to direct environmental policies toward their own goals, away from those of the average or median voter.

Anecdotal evidence of the persuasiveness of interest group activities is widespread, but there is relatively little systematic empirical evidence of the effectiveness of interest group informational campaigns. Chapters 6 and 7 of

---

21. See for example Stigler 1971; Niskanen 1971; Breton and Wintrobe 1975; Peltzman 1976; or Weingast and Moran 1983 for models of bureaucratic discretion. Weingast and Moran argue that congressional oversight committees may perfectly control bureaucracies through their control of bureau budgets.

22. Interest groups may also legally make strategic contributions to candidates and engage in single-issue voting. Interest groups may also use various forms of legal (and illegal) bribes to induce congressmen or regulators to adopt rules which advance interest group goals rather than the perceived goals of a typical voter. The recently regulated revolving door formerly allowed regulators to take lucrative jobs with those they had regulated in their agency jobs. Although I know of no statistical test of the effect of this reform, one expects that representative "shirking" would tend to diminish as a consequence.

23. Congleton and Sweetser (1992) demonstrate that even unbiased information can have deleterious effects insofar as increased knowledge of the distributional effects of policies may make even net benefit increasing policies more problematic to pass.

this volume provide statistical evidence that interest group *testimony* at public hearings has had significant effects on environmental and natural resource policy decisions made by environmental regulatory agencies.[24]

Of course, not all environmentally relevant information is generated by interest group activities. General scientific, aesthetic, and technological advances also tend to affect the public and private demand for environmental quality in several ways. First, technological advances tend to reduce the effective price of desired manufactured goods relative to environmental quality and lead to an increase in manufactured outputs.[25] Second, increased productivity implies that personal incomes rise, which tends to increase the demand for all normal goods, including environmental quality. Third, as knowledge and appreciation of natural phenomena increase, more subtle forms of environmental quality may be discerned and demanded.

Fourth, moreover, technological advances may directly and indirectly affect the balance of political power within a polity. For example, some forms of technological advance tend to decrease the marginal cost of organizing and sustaining interest group activities. Higher personal incomes together with reduced costs tend to increase the number of potential subscribers to policy advocacy services.[26] If contributions to interest groups supporting environmentally more stringent rules increase faster than for groups opposing greater stringency, then interest groups will tend to reinforce the increased popular demand for more environmental quality. If not, interest groups may reduce net pressures for more stringent environmental regulations. In either case, the increased effectiveness of resources invested in political activities tends to cause a more visible and intense political conflict over environmental policies as technology improves through time.[27]

---

24. It bears noting that observed policy changes tend to understate the overall effect of interest groups on the polity. Lobbying activities are not often one sided, and the efforts of opposing interest groups may largely offset each other, yielding little change in policy. For nearly every politically active group that gains relatively large benefits from a particular party there is another which bears large losses. In such cases, the efforts of each side may be highly effective, yet policies may be only modestly affected. Moreover, such lobbying efforts are a dead weight loss for society as a whole insofar as scarce economic resources have been devoted to jointly ineffectual activities. The same "result" could have been achieved at a lower cost had competing parties mutually reduced their efforts.

25. Technological advances reduce production costs (often by using resources in a manner that reduces effluent outputs), make whole new product areas and effluents possible, and/or induce changes in the methods used to appreciate environmental quality. The latter may increase overall effluent emissions even if effluents per unit of output have been reduced.

26. Congleton and Sweetser (1992) develop evidence that the extent and impact of interest group activities tends to increase with technological advances which reduce the cost of information production and dissemination.

27. It is, thus, not surprising that the most visible and intensely politicized debates over environmental policies occur in countries which make the most use of technologically sophisti-

An Overview of the Book: Analysis and Evidence

The chapters in this book analyze the political economy of specific environmental policy issues in detail. While no single contribution aims at the totality of environmental issues, together they shed considerable light on important links in the political chain which jointly determine environmental policies.

Interest Groups and Majoritarian Politics

The analysis above suggests that voter interests, interest group activities, and bureaucratic discretion all affect environmental policies. However, the direction of influence and the relative importance of these competing interests remains to be examined. Chapters 2 and 3 analyze the extent to which interest groups may determine the policy instruments used to address environmental problems. Chapter 4 attempts to determine empirically the extent to which campaign contributions by interest groups and legislator discretion affect environmental policy.

In chapter 2, James M. Buchanan and Gordon Tullock analyze the extent to which economic interest groups prefer standards over effluent charges as policy instruments for pollution management. Standards tend to diminish the profits of firms by less than equivalent Pigovian taxes. (Under standards firms only have to pay for the effluent-reducing equipment. Under a Pigovian tax, firms purchase similar effluent reducing equipment and generally pay a tax on their remaining effluent emissions.) Consequently economic interest group activities tend to support standards over what to most economists is the generally more desirable regulatory device, Pigovian taxes.

In chapter 3, Michael P. Leidy and Bernard M. Hoekman explore the extent to which international trade concerns may affect domestic and international environmental policies. They argue that firms in either a large or small open economy have even stronger reasons to lobby for inefficient environmental policy measures, output controls, than those identified by Buchanan and Tullock. They argue that inefficient environmental regimes strengthen the trade-policy linkage in several ways. For example, inefficient environmental regimes often establish precedents for market sharing which may be extended to foreign firms. These barriers to entry tend to increase profits over those which would have been associated with market competition under nonprotectionist trade policies.

In chapter 4, Dennis Coates analyzes the extent to which the campaign

cated production and communication methods. Nor is it surprising, for the same reasons, that the intensity of environmental debate is often greatest in areas where environmental quality is highest.

contributions of interest groups have had an impact on votes cast by representatives in the U.S. House of Representatives. The empirical work focuses on roll call votes on the Walker amendments to the Oregon and California bills designating federal lands as wilderness areas. These amendments gave the Secretary of Agriculture authority to waive provisions of the bill that increase unemployment. In the case of the Oregon proposal the Secretary of Agriculture could take unilateral action. In the case of the proposed amendment to the California bill, the Secretary of Agriculture could take such action only if requested by the state of California. Both amendments were voted down. Using ordered probit analysis, Coates analyzes the role that campaign contributions, representative characteristics, political party, the ideology of the representative's district, and characteristics of the electorate played in the representatives' votes. He finds that all these factors affected voting at the margin, although campaign contributions were not decisive.

The Environmental Bureaucracy

Once enacted, environmental legislation is implemented by standing organizations which are delegated authority by the legislature. How that discretion is used, and the extent to which it opens up regulatory processes to efforts by special interest groups, is analyzed in chapters 5, 6, and 7.

In chapter 5, George Van Houtven demonstrates that the EPA does not have a monolithic objective function. Different subunits of the EPA make use of their discretion in different ways. For example, the National Emissions Standards for Hazardous Air Pollutants (NESHAPs) grants the agency no explicit authority to take account of the costs and benefits of regulatory costs. The Toxic Substances Control Act (TSCA, or asbestos ban), explicitly directs the EPA to take account of costs and benefits when banning particular uses of asbestos and other toxics. In both cases, the EPA appears to have taken account of the costs and benefits of program regulations, but in a manner opposite to what might have been expected given the legislative mandates. Greater weight was given to net benefits under NESHAPs than under TSCA. In both cases, the EPA's reliance or nonreliance on cost-benefit analysis was challenged in the courts (in the former by environmental groups and in the latter by business interests). And in both cases, the EPA was found to have strayed too far from its legislative mandates.

In chapter 6, Maureen Cropper, William N. Evans, Stephen J. Berardi, Maria M. Ducla-Soares, and Paul Portney analyze the effects of cost-benefit analysis and interest group lobbying activities on EPA decisions to allow the continued use of various cancer-causing pesticides between 1975 and 1989. By law, the agency is encouraged to take account of economic costs and benefits. Their results reveal that agency officials did account for economic costs and benefits. However, their results reveal that the testimony of interest groups also affected the probability that a particular pesticide use could be continued. The testimony of environmental groups increased the probability that a pesticide use would be canceled. Testimony by grower organizations reduced the probability that a permitted use would be ended.

In chapter 7, Porter Hoagland and Scott Farrow examine the extent to which economic analysis and political interests have affected U.S. Department of Interior decisions regarding planned and actual sales of offshore drilling leases. The offshore leasing process consists essentially of two stages. In the first stage, tracts of undersea land are evaluated for their marine and geological properties, hearings are held, and a subset of the range of possible lease sites is announced. Subsequently, the original list of sites may be modified as a result of further analysis, pressure, and debate, as leases between the federal government and oil firms are consummated. Their analysis suggests that interest group activities have had significant effects on both stages of the leasing process, but generally have had stronger effects on the leases actually consummated than on the original planning list. In the final analysis, they could not reject a pure interest group model in which the efforts of oil companies and environmental groups completely determined the leasing outcome.

Relations between Governments

There are many cases in which a government interested in improving environmental quality within its own jurisdiction cannot do so on its own authority because local environmental quality is partly decided by regulations adopted by neighboring governments. In these cases, the problem of environmental regulation becomes similar to the original externality problem of coordinating private environmental protection.

In chapter 8, Gerald R. Faulhaber and Daniel E. Ingberman analyze properties of alternative institutional solutions to the local government "not in my backyard" (NIMBY) game. They examine three institutional arrangements for locating waste facilities: (1) market, (2) centralized democratic government, and (3) centralized democratic government with host veto. They find that the assignment of bargaining power is crucial to the efficiency of these alternative regimes. If information is observable and the potential host community has bargaining power, then either competitive or monopoly markets for the NIMBY activity (waste disposal) generates efficient location of facilities. Similar results are obtained for majoritarian central government. The strong conclusion of their analysis is that bargaining power rather than market versus government determines the efficiency of policy outcomes in the settings explored.

A related problem faced by firms is the extent to which governments will respond to locational decisions to locate environmentally risky facilities with new, more stringent environmental regulations. Most communities prefer that noxious, but necessary, waste dumps be located elsewhere, which gives rise to the NIMBY game among communities. In chapter 9, James T. Hamilton analyzes the locational choices of private waste disposal firms who are able to locate in several alternative communities. Once a site is chosen, however, firms bear a risk that communities will change their environmental regulations (or demand greater compensation) in response to the firm's locational choice. Hamilton demonstrates that the likelihood of political responses to locational choice is a prospective cost which influences the siting of NIMBYs. Statistical evidence supports his analysis in that a proxy for anticipated regulatory response is found to influence the locational choices of firms.

Environmental Treaties

The problems confronting national governments in addressing international environmental problems are similar to those of local governments. International environmental problems cannot be entirely (or in some cases even partly) controlled by any single government. International environmental problems differ from those of local governments, however, in that there exists neither a higher level of government from which broader regulations can be solicited, nor an effective enforcement system to enforce agreements reached between affected countries.

The most widely used solution to international environmental problems is a Coasian contract between interested governments, that is to say, an environmental treaty. In chapter 10, Werner W. Pommerehne provides a detailed examination of a specific instance of a local Coasian solution between two neighboring towns on the German/French border. The mayor of the German town of Kleinblittersdorf, fearing effluents from an upwind incinerator planned by the French community of Grosbliederstroff, encouraged his constituents to donate money for upgrading the French incinerator. Somewhat surprisingly, the mayor obtained sufficient contributions and loans for this purpose.

Of course, not every environmental agreement between governments will actually advance the environmental goals exposued. In chapter 11, Todd Sandler demonstrates that a subglobal carbon tax treaty may have little effect on atmospheric carbon oxide levels when nonsignatory nations react in an optimizing fashion. The analysis is cast in terms of carbon emissions and the current interest in prospects for global warming, but his results apply to many other forms of international air or water pollution. The analysis compares and contrasts the equilibrium policies that obtain under Nash and various forms of leader-follower equilibria. He argues that cooperative supranational agreements on the environment must be judged in light of the responses that nations outside the agreement may make.

In the end, domestic political institutions and interests determine whether a nation signs even imperfect environmental treaties. In chapter 12, I examine the extent to which elections affect the propensities of governments to regulate environmental matters. Democracies tend to be materially more prosperous nations than dictatorships. Were democracies not more inclined than dictatorships to regulate environmental matters, they would also tend to have more polluted environments than dictatorships. The paper demonstrates that the relative price of environmental protection for dictators is greater than that faced by median voters. Consequently, dictators may tend to adopt *less* stringent environmental rules than democracies in otherwise similar situations. Estimates of propensities to sign international treaties on the environment, specifically the Vienna Treaty and Montreal Protocol on CFC emissions, are consistent with this conclusion. Dictatorial countries were much less inclined to sign these two CFC agreements than were democracies.

Overall, the contributions to this book indicate that electoral constraints are not completely binding. Consequently, environmental policies reflect a variety of political and economic factors, including but not limited to special interest group activities. The statistical evidence developed suggests that neither cost-benefit considerations nor voter interests are entirely neglected in the development and implementation of environmental policies.[28]

Conclusion

Environmental politics may be thought of as a form of institutional production where new laws are created which induce individuals to alter the manner in which they use natural resources.[29] Productive environmental rules increase "value" for relevant decision makers by redirecting the use of resources in a manner which improves environmental quality sufficiently to offset perceived costs.

---

28. Not all of this discretion reduces welfare. The bureaucracy (fortunately) often takes greater account of costs and benefits or regulations than required, or permitted, by its enabling legislation.

29. Examples of informal codes of conduct include: Don't litter, recycle, buy and think green. A popular green bumper sticker around Washington area is "think globally, act locally." Such rules apply to a wide array of environmentally relevant private decisions.

Most formal legal regulations that are thought of as "environmental protection" deal with problems of waste disposal. That is to say, the laws restrict or enable the disposal of that subset of the products produced by manufacture which are themselves valueless or worse. However, it bears noting that these "environmental" problems are often partly the result of other policies which promote more extensive use of natural resources and/or create (implicit) rights to *freely* use common resources, such as the air or water, for purposes of waste disposal.

The value enhancing characteristics of such environmental policies is, as is the case of ordinary market production, at root a subjective matter, which may or may not have an objective counterpart. Individuals must determine whether they expect alternative policies to improve environmental quality from their own perspective. But there are rarely external criteria by which environmental judgments may be brought into complete accord. The pattern of environmental regulations and sanctions that we actually observe reflects the influence of political decision makers as constrained by the political institutions under which policies are adopted.

The various positive analyses of this volume shed indirect light on the manner in which policy proposals should be evaluated in light of political-economic concerns. The chapters by Faulhaber and Ingberman, Sandler, and Congleton suggest that in many cases institutions are the root of environmental problems. This contrasts with the usual economic approach to environmental problems in which political and legal institutions are ignored or taken as given. An implication of the analyses of this volume is that policy proposals should be based on a realistic assessment of the likely implementation of alternative rules under existing political and institutions. Proposed institutional reforms should analyze incentives for the exercise of undesirable and desirable discretion along the chain from voters to final environmental policies.[30] Not all institutional or rule changes are politically, legally, or behaviorally feasible. Would that we all simply did what was best in all circumstances.

In conclusion, it is worth remembering that substantial progress has been made on environmental issues through time in spite of the difficulty of the problems addressed. Organized human society has faced difficult political choices regarding environmental degradation and economic growth from the dawn of history. Ongoing communities living in more or less fixed locations have necessarily addressed and solved the problems of refuse disposal and water quality. Scientific advances have allowed once nearly intractable local problems to be readily addressed. More recently, as man's impact and wealth have increased, large-scale regional air and water quality problems have been addressed. While it cannot be claimed that economic development has been

---

30. As in ordinary production, the role of scientific information about the properties of alternative transformation processes is relevant to the assessment of the relative merits of alternative environmental rules. An understanding of the properties of natural human processes under existing rules is, in this sense, prior to any claim that existing rules are less productive than they might be. In cases where formal legal methods are to be employed, as opposed to persuasion or economic contracting, an understanding of political processes becomes central to any examination of existing environmentally relevant rules, and to any forecasts of the likely time path that such rules will follow in the future.

accomplished in the least costly manner, most would agree that air and water quality in developed countries is better today than it was forty years ago.

Now, even more ambitious environmental concerns have been raised, as large-scale biological systems and global climate have been added to the list of environmental necessities. However, the essential problem of managing environmental quality, although larger in scope, has not changed very much. In the end, whether public policies exacerbate or moderate the relationship between politics, pollution, and prosperity reflects the political balance between the interests of individual voters and the pressures of economic and environmental interest groups.

REFERENCES

Aristotle. 1969. *The Politics.* Trans. T. A. Sinclair. Baltimore: Penguin Books.
Black, D. 1958. *The Theory of Committees and Elections.* New York: Cambridge University Press.
Becker, G. 1983. "A Theory of Competition among Pressure Groups for Political Influence." *Quarterly Journal of Economics* 98:371–400.
Block, W. E. 1990. *Economics and the Environment: A Reconciliation.* Vancouver: Frazier Institute.
Breton, A., and Wintrobe, R. 1975. "The Equilibrium Size of a Budget Maximizing Bureau." *Journal of Political Economy* 83:195–207.
Buchanan, J. M. 1969. *Cost and Choice.* Chicago: Markham Publishing Company.
Buchanan, J. M.; Tollison, R. D.; and Tullock, G., eds. 1980. *Towards a Theory of the Rent-Seeking Society.* College Station: Texas A&M Press.
Cleeper, H., ed. 1966. *Origins of American Conservation.* New York: Ronald Press Company.
Congleton, R. D. 1982. "A Model of Asymmetric Bureaucratic Inertia and Bias." *Public Choice* 39:421–25.
Congleton, R. D., and Shughart, W. F., II. 1990. "The Growth of Social Security: Electoral Push or Political Pull?" *Economic Inquiry* 27:109–32.
Congleton, R. D., and Sweetser, W. 1992. "Political Deadlocks and Distributional Information: The Value of the Veil." *Public Choice* 73:1–19.
Denzau, A., and Grier, K. 1984. "Determinations of Local School Spending: Some Consistent Estimates." *Public Choice* 44:375–83.
Dryzek, J. S. 1987. *Rational Ecology: The Environment and Political Economy.* New York: Basil Blackwell.
Enelow, J. M., and Hinich, M. J. 1984. *The Spatial Theory of Voting.* Cambridge: Cambridge University Press.
Goodin, R. E. 1992. *Green Political Theory.* Cambridge: Polity Press.
Hardin, G. 1968. "The Tragedy of the Commons." Science 162:1243–48.
Holcombe, R. G. 1980. "An Empirical Test of the Median Voter Model." *Economic Inquiry* 17:260–74.

Kalt, J. P., and Zupan, M. A. 1984. "Capture and Ideology in the Theory of Politics." *American Economic Review* 74:279–300.

Macfarlane, A. 1978. *The Origins of English Individualism: The Family, Property, and Social Transition.* New York: Cambridge University Press.

Meyer, A. B. 1966. "Forests and Forestry." In H. Cleeper, ed., *Origins of American Conservation,* 38–56. New York: Ronald Press Company.

Mueller, D. 1989. *Public Choice II.* New York: Cambridge University Press.

Nash, R., ed. 1968. *The American Environment: Readings in the History of Conservation.* Reading, Mass: Addison Wesley.

Niskanen, W. A. 1971. *Bureaucracy and Representative Government.* Chicago: Aldine Atherton.

Ostrom, E. 1990. *Governing the Commons: The Evolution of Institutions for Collective Action.* New York: Cambridge University Press.

Pashigan, B. Peter. 1985. "Environmental Regulation: Whose Self-Interests are Being Protected?" *Economic Inquiry* 23:551–84.

Peltzman, S. 1976. "Toward a More General Theory of Regulation." *Journal of Law and Economics* 19:211–40.

Shughart, W. F., II; Tollison, R. D.; and Goff, B. L. 1986. "Bureaucratic Structure and Congressional Control." *Southern Economic Journal* 52:962–72.

Sopper, W. E. 1966. "Watershed Management." In H. Cleeper, ed., *Origins of American Conservation,* 101–18. New York: Ronald Press Company.

Stigler, G. J. 1971. "The Theory of Economic Regulation." *Bell Journal of Economics and Management Science* 13:1–10.

Weingast, B. R., and Moran, M. J. 1983. "Bureaucratic Discretion or Congressional Control? Regulatory Policymaking by the Federal Trade Commission." *Journal of Political Economy* 91:765–800.

Part 2
Interest Groups and
Majoritarian Politics

CHAPTER 2

Polluters' Profits and Political Response: Direct Controls versus Taxes

James M. Buchanan and Gordon Tullock

Economists of divergent political persuasions agree on the superior efficacy of penalty taxes as instruments for controlling significant external diseconomies which involve the interaction of many parties. However, political leaders and bureaucratic administrators, charged with doing something about these problems, appear to favor direct controls. Our purpose in this paper is to present a positive theory of externality control that explains the observed frequency of direct regulation as opposed to penalty taxes or charges. In the public-choice theory of policy,[1] the interests of those who are subjected to the control instruments must be taken into account as well as the interests of those affected by the external diseconomies. As we develop this theory of policy, we shall also emphasize an elementary efficiency basis for preferring taxes and charges which heretofore has been neglected by economists.

I

Consider a competitive industry in long-run equilibrium, one that is composed of a large number of *n* identical producing firms. There are no productive inputs specific to this industry, which itself is sufficiently small relative to the economy to insure that the long-run supply curve is horizontal. Expansions and contractions in demand for the product invoke changes in the number of firms, each one of which returns to the same least-cost position after adjustment. Assume that, from this initial position, knowledge is discovered which indicates that the industry's product creates an undesirable environmental side effect. This external diseconomy is directly related to output, and we assume there is no technology available that will allow alternative means of producing

Reprinted from *American Economic Review* 65, no. 1 (1975): 139–47. We wish to thank the National Science Foundation for research support. Needless to say, the opinions expressed are our own.

1. Charles Goetz imposes a public choice framework on externality control, but his analysis is limited to the determination of quantity under the penalty-tax alternative.

the private good without the accompanying public bad. We further assume that the external damage function is linear with respect to industry output; the same quantity of public bad per unit of private good is generated regardless of quantity.[2] We assume that this damage can be measured and monitored with accuracy.

This setting has been deliberately idealized for the application of a penalty tax or surcharge. By assessing a tax (which can be computed with accuracy) per unit of output on all firms in the industry, the government can insure that profit-maximizing decisions lead to a new and lower industry output that is Pareto optimal. In the short run, firms will undergo losses. In the long run, firms will leave the industry and a new equilibrium will be reached when remaining firms are again making normal returns on investment. The price of the product to consumers will have gone up by the full amount of the penalty tax.

No one could dispute the efficacy of the tax in attaining the efficient solution, but we should note that in this setting, the same result would seem to be equally well insured by direct regulation. Policymakers with knowledge of individual demand functions, the production functions for firms and for the industry, and external damage functions could readily compute and specify the Pareto-efficient quantity of industry output.[3] Since all firms are identical in the extreme model considered here, the policymakers could simply assign to each firm a determinate share in the targeted industry output. This would then require that each firm reduce its own rate of output by X percent, that indicated by the difference between its initial equilibrium output and that output which is allocated under the socially efficient industry regulation.[4]

Few of the standard arguments for the penalty tax apply in this setting. These arguments have been concentrated on the difficulties in defining an efficient industry output in addition to measuring external damages and on the difficulty in securing data about firm and industry production and cost functions. With accurately measured damage, an appropriate tax will insure an efficient solution without requiring that this solution itself be independently computed. Or, under a target or standards approach, a total quantity may be computed, and a tax may be chosen as the device to achieve this in the absence of knowledge about the production functions of firms.[5]

In the full information model, none of these arguments is applicable. There is, however, an important economic basis for favoring the penalty tax

2. This assumption simplifies the means of imposing a corrective tax. For some of the complexities, see Otto Davis and Andrew Whinston and Stanislaw Wellisz.

3. See Allen Kneese and Blair Bower, p. 135.

4. No problems are created by dropping the assumption that firms are identical so long as we retain the assumption that production functions are known to the regulator.

5. This is the approach taken by William Baumol, who proposes that a target level of output be selected and a tax used to insure the attainment of this target in an efficient manner.

over the direct control instrument, one that has been neglected by economists. The penalty tax remains the preferred instrument on strict efficiency grounds, but, perhaps more significantly, it will also facilitate the enforcement of results once they are computed.[6] Under the appropriately chosen penalty tax, firms attain equilibrium only at the efficient quantity of industry output. Each firm that remains in the industry after the imposition of the tax attains long-run adjustment at the lowest point on its average cost curve only after a sufficient number of firms have left the industry. At this equilibrium, there is no incentive for any firm to modify its rate of output in the short run by varying the rate of use of plant or to vary output in the long run by changing firm size. There is no incentive for resources to enter or to exit from the industry. So long as the tax is collected, there is relatively little policing required.

This orthodox price theory paradigm enables the differences between the penalty-tax instrument and direct regulation to be seen clearly. Suppose that, instead of levying the ideal penalty tax, the fully informed policymakers choose to direct all firms in the initial competitive equilibrium to reduce output to the assigned levels required to attain the targeted efficiency goal for the industry. No tax is levied. Consider figure 1, which depicts the situation for the individual firm. The initial competitive equilibrium is attained when each firm produces an output, q_i. Under regulation it is directed to produce only q_0, but no tax is levied. At output q_0, with an unchanged number of firms, price is above marginal cost (for example price is at P'). Therefore, the firm is not in short-run equilibrium and would, if it could, expand output within the confines of its existing plant. More importantly, although each firm will be producing the output quota assigned to it at a somewhat higher cost than required for efficiency reasons, there may still be an incentive for resources to enter the industry. The administrator faces a policing task that is dimensionally different from that under the tax. He must insure that individual firms do not violate the quotas assigned, and he must somehow prevent new entrants. To the extent that the administrator fails in either of these tasks, the results aimed for will not be obtained. Output quotas will be exceeded, and the targeted level of industry production overreached.

If the administrator assigns enforceable quotas to existing firms and successfully prevents entrants, the targeted industry results may be attained, but there may remain efficiency loss since the industry output will be produced at higher average cost than necessary if firms face U-shaped long-run average cost curves. Ideally, regulation may have to be accompanied by the assignment of full production quotas to a selected number of the initial firms in the industry. This policy will keep these favored firms in marginal adjust-

---

6. See George Hay. His discussion of the comparison of import quotas and tariffs on oil raises several issues that are closely related to those treated in this paper.

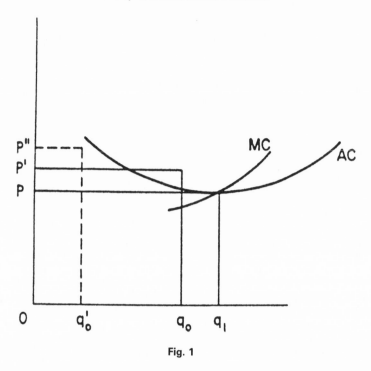

Fig. 1

ment with no incentives for in-firm adjustments that might defeat the purpose of the regulation. But even more than under general quota assignment there will be strong incentives for firms to enter the industry and to secure at least some share of the rents that the restriction of industry output generates. If the response to this pressure should be that of reassigning quota shares within the unchanging and targeted industry output so as to allow all potential entrants some share, while keeping all firms, actual and potential, on an equal quota basis, the final result may be equivalent to the familiar cartel equilibrium. No firm will be earning more than normal returns, but the industry will be characterized by too many firms, each of which produces its assigned output inefficiently.

II

When we examine the behavioral adjustments to the policy instruments in the manner sketched out above, a theory of policy emerges. Regulation is less desirable on efficiency grounds even in the presence of full information, but this instrument will be preferred by those whose behavior is to be subjected to either one or the other of the two policy instruments. Consider the position of the single firm in the fully competitive industry, depicted in figure 1. Under the imposition of the tax, short-run losses are necessarily incurred, and the firm reattains normal returns only after a sufficient number of its competitors have shifted resources to other industries. The tax reduces the present value of the firm's potential earnings stream, whether the particular firm remains in the industry after adjustment or withdraws its investment and shifts to alternative employment. In terms of their own private interests, owners of firms in the industry along with employees will oppose the tax. By contrast, under regulation firms may well secure pecuniary gains from the imposition of direct controls that reduce total industry output. To the extent that the restriction is achieved by the assignment of production quotas to existing firms, net profits may be present even for the short term and are more likely to arise after adjustments in plant. In effect, regulation in this sense is the directional equivalent of cartel formation provided that the individual firm's assigned quota falls within the limited range over which average cost falls below price. Such a range must, of course, exist, but regulatory constraints may possibly be severe enough to shift firms into positions where short-term, and even possibly long-term, losses are present, despite increased output price. Such a result is depicted by a restriction to q_0' in figure 1, with price at P''.

Despite the motivation which each firm has to violate assigned quotas under regulation, it remains in the interest of firms to seek regulatory policy that will enforce the quotas. If existing firms foresee the difficulty of restricting entry, and if they predict that governmental policymakers will be required to accommodate all entrants, the incentive to support restriction by regulation remains even if its force is somewhat lower. In final cartel equilibrium, all the firms will be making no more than normal returns. But during the adjustment to this equilibrium, above-normal returns may well be available to all firms that hold production quotas. Even if severe restriction forces short-term losses on firms, these losses will be less than those under the tax. Rents over this period may well be positive, and even if negative, they will be less negative than those suffered under the tax alternative. Therefore, producing firms will always oppose any imposition of a penalty tax. However, they may well favor direct regulation restricting industry output, even if no consideration at all is given to the imposition of a tax. And, when faced with an either/or choice, they will always prefer regulation to the tax.

III

There is a difference between the two idealized solutions that has not yet been discussed, and when this is recognized, the basis of a positive hypothesis about policy choice may appear to vanish. Allocationally, direct regulation can produce results equivalent to the penalty tax, providing that we neglect enforcement cost differentials. *Distributionally,* however, the results differ. The imposition of tax means that government collects revenues (save in the case where tax rates are prohibitive) and these must be spent. Those who anticipate benefits from the utilization of tax revenues, whether from the provision of publicly supplied goods or from the reduction in other tax levies, should prefer the tax alternative, and they should make this preference known in the political process. To the extent that the beneficiaries include all or substantially all members of the community, the penalty tax should carry the day. Politicians, in responding to citizenry pressures, should heed the larger number of beneficiaries and not the disgruntled members of one particular industry. This political choice setting is, however, the familiar one in which a small, concentrated, identifiable, and intensely interested pressure group may exert more influence on political choice making than the much larger majority of persons, each of whom might expect to secure benefits in the second order of smalls.

There is an additional reason for predicting this result with respect to an innovatory policy of externality control. The penalty tax amounts to a legislated change in property rights, and as such it will be viewed as confiscatory by owners and employees in the affected industry. Legislative bodies, even if they operate formally on majoritarian principles, may be reluctant to impose what seems to be punitive taxation. When, therefore, the regulation alternative to the penalty tax is known to exist, and when representatives of the affected industry are observed strongly to prefer this alternative, the temptation placed on the legislator to choose the direct control policy may be overwhelming, even if he is an economic theorist and a good one. Widely accepted ethical norms may support this stance; imposed destruction of property values may suggest the justice of compensation.[7]

If policy alternatives should be conceived in a genuine Wicksellian framework, the political economist might still expect that the superior penalty tax should command support. If the economist ties his recommendation for the penalty tax to an accompanying return of tax revenues to those in the industry who suffer potential capital losses, he might be more successful than he has been in proposing unilateral or one-sided application of policy norms. If revenues are used to subsidize those in the industry subjected to capital losses from the tax, and if these subsidies are unrelated to rates of output, a two-sided tax subsidy arrangement can remove the industry source of opposition while still insuring efficient results. In this respect, however, economists themselves have failed to pass muster. Relatively few modern economists who have engaged in policy advocacy have been willing to accept the Wicksellian

---

7. For a comprehensive discussion of just compensation, see Frank Michelman.

methodological framework which does, of course, require that some putative legitimacy be assigned to rights existent in the status quo.[8]

IV

To this point we have developed a theory of policy for product-generated external diseconomies, the setting which potentially counterposes the interest of members of a single producing industry against substantially all persons in the community. External diseconomies may, however, arise in consumption rather than in production, and these may be general. For purposes of analysis, we may assume that all persons find themselves in a situation of reciprocal external diseconomies. Traffic congestion may be a familiar case in point.

The question is one of determining whether or not persons in this sort of interaction, acting through the political processes of the community, will impose on *themselves* either a penalty tax or direct regulation. We retain the full information assumption introduced in the production externality model. For simplicity here, consider a two-person model in which each person consumes the same quantity of good or carries out the same quantity of activity in the precontrol equilibrium, but in which demand elasticities differ. Figure 2 depicts the initial equilibrium at E with each person consuming quantity Q. The existence of the reciprocal external diseconomy is discovered. The community may impose an accurately measured penalty tax in the amount T, in which case A will reduce consumption to Q_a and B will reduce consumption to Q_b. Total consumption is reduced from $2Q$ to $(Q_a + Q_b)$, but both A and B remain in equilibrium. At the new price P', which includes tax, neither person desires to consume more or less than the indicated quantities. The government collects tax revenues in the amount $[2(PP'JH) + HJLA]$. Alternatively, the community may simply assign a restricted quantity quota to each person. If the government possesses full information about demand functions, it can reduce A's quota to Q_a and B's quota to Q_b, securing results that are allocatively identical to those secured by the tax. However, under the quota, both A and B will find themselves out of equilibrium; both will, if allowed quantity adjustment, prefer to expand their rate of consumption.

It will be useful to examine the ideal tax against the quota scheme outlined above, which we may call the idealized quota scheme. If individuals expect no returns at all from tax revenues in the form of cash subsidies, public goods benefits, or reductions in other taxes, both A and B will clearly prefer the direct regulation. The loss in consumers' surplus under this alternative is small relative to that which would be lost under the penalty tax. Each person

---

8. For a specific discussion of the Wicksellian approach, see Buchanan (1959).

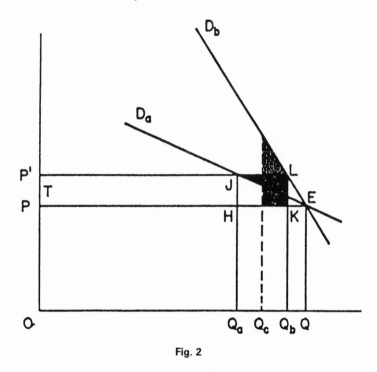

Fig. 2

willingly trades off marginal quantity adjustment for the more favorable infra-marginal terms offered under direct regulation, given our assumptions that both instruments achieve the same overall externality control objective.

Under extreme fiscal illusion, individuals may ignore benefits from tax revenues, but consistent methodological precept requires that we allow persons to recognize the benefit side of the fiscal account, at least to some degree. Let us allow all revenues under the penalty tax to be returned in equal shares to all taxpayers. Under this arrangement, each person expects to get back one-half of the amount measured as indicated above for figure 2. Simplifying, each expects to get back the amount *PP'JH*, which he personally pays in, plus one-half of the amount measured by the rectangle *JHKL*, all of which is paid in by *B*. From an examination of figure 2, it is clear that individual *A* will favor the penalty tax under these assumptions. The situation for individual *B* is different; he will prefer direct regulation. He will secure a differential gain measured by the horizontally shaded area in figure 2, which is equal to the differential loss that individual *A* will suffer under this alternative. The policy result, insofar as it is influenced by the two parties, is a standoff under this idealized tax and idealized quota system comparison.

For constitutional and other reasons, control institutions operating within a democratic order could scarcely embody disproportionate quota assignments. A more plausible regulation alternative would assign quotas proportionate to initial rates of consumption, designed to reduce overall consumption to the level indicated by target criteria. The comparison of this alternative with the ideal tax arrangement is facilitated by the construction of figure 2 where the initial rates of consumption are equal. In this new scheme, each person is assigned a quota Q_c, which he is allowed to purchase at the initial price P. We want to compare this arrangement with the ideal tax, again under the assumption that revenues are fully returned in equal per head subsidies. As in the first scheme, both persons are in disequilibrium at quantity Q_c and price P. The difference between this model and the idealized quota scheme lies in the fact that at Q_c, the marginal evaluations differ as between the two persons. There are unexploited gains from trade, even under the determined overall quantity restriction.

It will be mutually advantageous for the two persons to exchange quotas and money, but, at this point, we assume that such exchanges do not take place, either because they are prohibited or because transactions costs are too high. Individual A will continue to favor the tax alternative but his differential gains will be smaller than under the idealized quota scheme. In the model now considered, A's differential gains under the ideal tax are measured by the blacked-in triangle in figure 2. Individual B may or may not favor the quota, as in the earlier model. His choice as between the two alternatives, the ideal tax on the one hand and the restriction to Q_c at price P on the other, will depend on the comparative sizes of the two areas shown as horizontally and vertically shaded in figure 2. As drawn, he will tend to favor the quota scheme, but it is clearly possible that the triangular area could exceed the rectangular one if B's demand curve is sufficiently steep in slope. In any case, the choice alternatives for both persons are less different in the net than those represented by the ideal tax and the idealized quota.

While holding all of the remaining assumptions of the model, we now drop the assumption that no exchange of quotas takes place between A and B. To facilitate the geometrical illustration, figure 3 essentially blows up the relevant part of figure 2. With each party initially assigned a consumption quota of Q_c, individual A will be willing to sell units to individual B for any price above his marginal evaluation. Hence, the lowest possible supply price schedule that individual B confronts is that shown by the line RL in figure 3. The maximum price that individual B is willing to pay for additional units of quota is his marginal evaluation, shown by SL. The gains-from-trade are measured by the triangular area RLS. The distribution of these gains will, of course, be settled in the strict two-man setting by relative bargaining skills, but let us assume that individual B, the buyer, wants to purchase consumption quota units from A, but also to do so in such a way that individual A will come

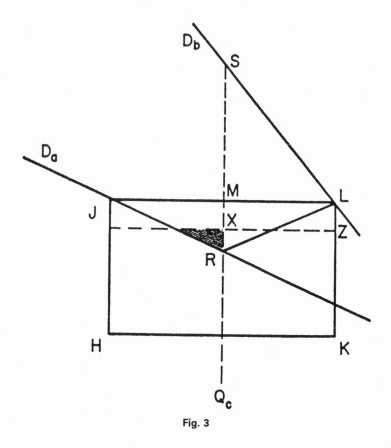

Fig. 3

to prefer this system over the tax. To accomplish this, he must insure that A gets a share of the net gains at least equal to the area RML on figure 3. Individual B, the buyer, retains gains of MSL under this division of the spoils. But in this arrangement, both persons are indifferent as between the policy alternatives. The system is on the Pareto frontier, and the quota scheme plus the exchange process produces allocative and distributive results identical to those generated under the ideal tax. This becomes the analogue of the Coase theorem in the context that we are examining.[9]

V

These somewhat inconclusive results may seem to provide anything but a positive theory of policy akin to that presented with respect to production externalities. The comparisons are, however, a necessary stage in developing

9. See Ronald Coase. For a related extension of the Coase theorem, see Buchanan (1973).

such a theory. Recall that we have made these comparisons under the most favorable possible assumption concerning anticipated return of revenues under the penalty tax. In the real world, individuals will not anticipate that these will be returned dollar-for-dollar, and they will tend to place at least some discount on the value of benefits that they expect.

Let us say that each person expects an aggregate benefit value of only 80 cents on the dollar from tax revenues collected under the penalty tax. Consider what this single change does to the results of the last comparison made, that which involves proportionate quota assignments along with a free market in quotas. In this case, individual *B*, the buyer, can offer individual *A*, the seller, more than the amount required to make him prefer the quota alternative, while himself continuing to secure differential benefit under this alternative. Individual *A*'s differential gains from the ideal penalty tax are reduced to the shaded area in figure 3. By paying individual *A* the amount measured by *RML*, he has improved *A*'s position relative to the penalty tax. And, in the process, he has retained for himself a differential gain measured by the area *MXZL*. Both persons in full knowledge of the alternatives will prefer the quota system, and political leaders will presumably respond by opting for regulation.

The same reasoning can readily be extended to apply to any quota system. In the idealized quota assignment first considered, we demonstrated that one person would favor the penalty tax and the other the quota. Individual *A*, who favors the penalty tax, loses no consumer's surplus, and he does expect to secure an income transfer through the return of tax revenues. When we modify the assumptions concerning expectations of the value of returned revenues or benefits, however, this conclusion need not hold. Individual *A* will, of course, expect to get back in benefits some part of the tax revenues paid in by *B* that is in excess of that contributed by *A* himself. If, however, individual *A* applies the same discount factor to all revenues collected, the deadweight loss may more than offset the income transfer effect. Examination of figure 2 indicates that under the 80 percent assumption, one-fifth of the area measured by *PP'JH* will represent deadweight loss to *A* from the revenues that he pays in. This deadweight loss may well be larger than the measure of the income transfer that he expects, which amounts to 80 percent of the horizontally shaded area in figure 2. Once we introduce any plausible discount factor into the expectation of individuals concerning the return of tax revenues, it is relatively easy to demonstrate situations under which both persons may be led by private self-interest to favor the direct regulation alternative.

VI

We have developed a positive theory of externality control policy for both the production and consumption interactions under highly abstract and simplified models which allows us to isolate influences on policy formation which have been neglected. Decisions on the alternative policy instruments in democratic governments are surely influenced by the preferences of those who are subjected to them. The public-choice approach, which concentrates attention on the individual's choice as between policy instruments, allows us to construct hypotheses that explain the prevalence of direct regulation.[10] For economists who continue to support the penalty tax alternative, the analysis suggests that they had best become good Wicksellians and begin to search out and invent institutional arrangements that will make the penalty tax acceptable to those who are primarily affected.

REFERENCES

W. J. Baumol. "On Taxation and the Control of Externalities." *Amer. Econ. Rev.* 62 (June 1972): 307–22.

J. M. Buchanan. "Positive Economics, Welfare Economics, and Political Economy." *J. Law Econ.* 2 (Oct. 1959): 124–38.

———. "The Coase Theorem and the Theory of the State." *Natur. Resources J.* 13 (Oct. 1973): 579–94.

——— and N. Tideman. "Gasoline Rationing and Market Pricing: Public Choice in Political Democracy." Research pap. no. 808231-1-12. Center for Study of Public Choice, Virginia Polytechnic Inst. and State Univ., Jan. 1974.

R. H. Coase. "The Problem of Social Cost." *J. Law Econ.* 3 (Oct. 1960): 1–44.

O. A. Davis and A. Whinston. "Externalities, Welfare, and the Theory of Games." *J. Polit. Econ.* 70 (June 1962): 241–62.

C. J. Goetz. "Political Equilibrium vs. Economic Efficiency in Effluent Pricing." In J. R. Conner and E. Loehman, eds., *Economic Decisionmaking for Environmental Control.* Gainesville, 1973.

G. A. Hay. "Import Controls on Foreign Oil: Tariff or Quota?" *Amer. Econ. Rev.* 61 (Sept. 1971): 688–91.

A. V. Kneese and B. T. Bower. *Managing Water Quality: Economics, Technology, Institutions.* Baltimore, 1968.

F. J. Michelman. "Property Utility, and Fairness: Comments on the Ethical Foundations of 'Just Compensation' Law." *Harvard Law Rev.* 80 (Apr. 1967): 1165–1257.

S. Wellisz. "On External Diseconomies and the Government-Assisted Invisible Hand." *Economica* 31 (Nov. 1964): 345–62.

---

10. Much of the analysis developed in this paper can be applied more or less directly to policy alternatives proposed in the energy crisis of late 1973 and early 1974. For such application, see Buchanan and Nicolaus Tideman.

CHAPTER 3

Pollution Abatement, Interest Groups, and Contingent Trade Policies

Michael P. Leidy and Bernard M. Hoekman

1. Introduction

In the last twenty years or so two seemingly unrelated policy developments have swept across a number of industrialized economies. First, there has been a dramatic rise in instances of sector-specific administered protection from foreign competition.[1] At the same time, environmental concerns have stimulated a policy response that continues to spread a patchwork of environmental rules and regulations across these same economies. When one looks at these environmental policies, efficient regimes have been largely rejected in favor of various inferior (second- or Nth-best) options. Both developments imply disturbing deadweight losses. They also warrant asking whether there are interactions between trade policy and environmental policy reinforcing these developments.

The purpose of this chapter is to analyze the political economy of environmental-policy formation in a trading economy with established rules for administered protection. It extends the closed-economy analysis of Buchanan and Tullock (1975), Maloney and McCormick (1982), and Yandle (1989), who have observed that there may be incentives for industrial polluters to support inefficient pollution-abatement policies. These authors point

We thank Kym Anderson, Richard Blackhurst, Peter Lloyd, Richard Snape, John Whalley, and Alan Winters for helpful comments on an early draft of this paper, which was written while the authors were with the Economic Research and Analysis Unit of the GATT Secretariat. The paper extends the analysis presented in Leidy and Hoekman 1994. The views expressed are our own and should not be attributed to the International Monetary Fund or the World Bank.

1. Administered protection, also called contingent protection, includes antidumping and countervailing duty procedures and escape clause mechanisms implementing Article XIX of the GATT (the *safeguards clause*). In the United States, for example, there were just three findings of dumping in the period 1952–60 and eleven during 1961–68, whereas during the period 1980–88 there were 385 antidumping cases of which 72 percent ended in restrictive outcomes, including price undertakings and negotiated restraint agreements (Finger and Murray 1990).

out that because industrial pollution control may imply restrictions on output, the opportunity arises for polluters to consolidate market power on the road to pollution control. We argue that the social costs associated with the adoption of an inefficient environmental regime are likely to be compounded by consequent restrictions on trade when the affected industries are import competing. The preferences of interest groups for alternative environmental regimes tend to be linked to the legal-institutional setting in which trade policy is conducted. Under existing rules and practices in the area of administered protection, there is reason to believe that interest group preferences for an inefficient approach to pollution control will be strengthened because the adoption of such a regime is more likely to imply a concomitant increase in trade barriers. It is shown that there may be a confluence of interests among import-competing polluters, environmental interests, labor groups, and even foreign exporters, all favoring an inefficient regulatory package. And this support derives in part from the heightened expectation of trade restrictions likely to accompany the inefficient environmental regime.

How might an inefficient environmental regime enhance the prospect of protection under existing administrative rules? It may do so by inducing structural adjustments that increase the likelihood of satisfying the injury criteria for protection; by setting up environmentally based barriers to entry that can help secure the profits of protection, thereby inducing more petitions than otherwise; by setting a precedent for market sharing that may facilitate the negotiation of a voluntary export restraint agreement (VER); and by establishing an Olsonian "other purpose" enabling the industry to speak more readily with one voice when petitioning for protection.

A sense of the significance of this analysis is suggested by an empirical study by Tobey (1990), who concluded that "the stringent environmental regulations imposed on industries in the late 1960s and early 1970s by most industrialized countries have not measurably affected international trade patterns in the most polluting countries" (p. 192). Three of the five industries included in his study, however, are heavily protected industries in most industrialized countries (primary iron and steel, chemicals, and paper and pulp). Our analysis suggests that the effects on trade patterns expected by Tobey (and others) need not emerge because new trade barriers which tend to offset such effects may be induced by environmental policy. Thus Tobey may be measuring the status-quo preserving effects of endogenous protection rather than the trade altering effects of pollution-control policy *ceteris paribus*. Indeed he speculates in his final paragraph that impediments to trade may be confounding his results.

Industries facing the highest pollution-abatement costs are among those most frequently seeking and receiving protection in industrialized countries. Metals, including basic metal products, and chemical products account for ten of the top 19 (of 122) U.S. industries ranked by pollution-abatement costs.[2] Industries in these categories accounted for 68 percent of all antidumping (AD) investigations (260 of 381 cases) and 78 percent of all definitive AD duties (119 of 153 actions) taken in the United States from July 1980 through June 1989 (GATT 1990). Among all the arrangements restraining exports to the United States in effect in 1989 (including voluntary export restraints, orderly marketing arrangements, industry-to-industry arrangements, and other VER-like arrangements), the subcategory of steel and steel products alone accounted for 51 percent (33 arrangements), by far more than any other single category (GATT 1989). A similar concentration of protectionist events in industries with high abatement costs appears in the European Community (EC) and Australia, two other major users of AD.[3] In the EC, the chemical industry has been the single most active user of AD law. It accounted for almost 40 percent of all AD cases initiated and 46 percent of those ending in a restrictive outcome between 1980 and 1987. When iron and steel is added, these sectors accounted for 56 percent of all restrictive outcomes from EC AD petitions during this period (GATT 1991). In Australia, petitions from basic metals and chemicals producers accounted for 45 percent of all AD investigations initiated between 1983 and 1989 (Banks 1990).

The following two examples are suggestive of the possible linkages between environmental regulation and administered protection. The Australian chemical industry has explicitly linked domestic environmental policy to the sharp increase in imports during 1989–90 and in a recent report warned that the emphasis on environmental issues "has reached unhealthy levels and may result in the substantial de-industrialization of Australia."[4] The chemical industry is the single most active initiator of AD petitions in Australia, having initiated one-third of all cases (91 in total) between 1983 and 1990 (Banks 1990). Following the 1989–90 surge in imports, between mid-1990 and mid-1991, 42 new AD cases were brought by the industry. This was over 60 percent of the total, a marked increase on its average share of one-third noted above.

The cement industry ranks first in pollution abatement costs among 122 U.S. industries. Relying predominantly on coal-fired ovens and kilns (USITC 1989), it has been affected by relatively stringent air-quality standards under the Clean Air Act. A 1977 amendment to the Clean Air Act established

---

2. Three-digit SIC rankings of industries by pollution-abatement costs per unit of output are reported in Low 1992. The original source is the U.S. Dept. of Commerce, *Manufacturers Pollution Abatement Capital Expenditures and Operating Costs,* Annual Survey of Manufacturers (1988).

3. The United States, the European Community, and Australia were the top three initiators of AD investigations during the 1980s (Finger 1987).

4. *Journal of Commerce,* February 4, 1991.

mandatory use of so-called scrubbing technology, regardless of levels of sulphur dioxide emissions, type of coal used, or local air quality. Since 1978 the U.S. industry (including regional associations) has petitioned six times for AD relief. The most recent petition was presented in 1989 by an association of southern-state producers. It resulted in the imposition, in August 1990,[5] of definitive duties on cement imports from Mexico, where environmental regulations are less strict.[6]

The analysis below provides a theoretical framework that identifies several dimensions of the suspected endogeneity of trade barriers to environmental regulations and how this endogeneity may influence interest-group preferences for alternative environmental policies. It suggests that the state of environmental regulations may be a significant new explanatory variable in commercial-policy analysis. In the next section, we consider the interests of four groups: domestic environmentalists, domestic polluters/producers, domestic sector-specific factors (labor), and foreign exporters. Whether a country is small or large, the polluting import-competing sector, domestic environmentalists, domestic labor interests, and foreign exporters are all likely to prefer the inefficient production regulation to the efficient penalty tax in the home economy. Existing commercial policy rules and practices in the area of contingent protection, particularly antidumping and voluntary restraint agreements, contribute to the scope and strength of this support. Section 3 considers ways in which policies might be formulated to undermine the coalition favoring the inefficient regulations. Section 4 concludes.

2. Penalty Taxes versus Quantity Regulation in a Trading Economy

Buchanan and Tullock (1975) examined the case of a polluting industry under perfect competition in a closed-economy setting. They showed that while a penalty tax is the efficient instrument to achieve any given level of pollution abatement, firms will tend to prefer quantity regulation because it may confer cartel-like gains. Maloney and McCormick (1982) reach a similar conclusion in a model that imposes a standards-based approach to pollution control. In this section the Buchanan-Tullock model is extended by incorporating a foreign exporting sector and explicitly considering the interests of environmen-

5. A petition was filed by the Ad Hoc Committee of Arizona-New Mexico-Texas-Florida Producers of Gray Portland Cement in September 1989. On May 18, 1990, the Ad Hoc Committee of Southern California Producers of Gray Portland Cement filed a separate petition against Japanese producers.

6. Mexico supplied half of all imports in the southern region at the time of the petition. It can also be noted that Mexican plants use oil-fired ovens and kilns, and therefore may be friendlier environmentally than U.S. coal-fired plants.

talists, foreign exporters, labor groups, and domestic producers of an import-competing good. One should not interpret our modeling choice as indicating that we believe the world of pollution control is restricted to a choice between penalty taxes or quantity regulation. This is clearly not the case.[7] Instead, this model offers a simple example of the way in which pressure groups will often block the efficient approach to pollution control. It also provides insight into the broad scope for unintended and socially costly linkages between environmental and commercial policies.

2.1. The Small-Country Case

Consider a competitive import-competing industry in long-run equilibrium. Assume that there are N_0 identical domestic firms in the initial equilibrium. Furthermore, assume that there are industry-wide pecuniary diseconomies of scale due to rising factor prices with industry expansion. This simply implies an upward sloping long-run market supply curve for domestic suppliers.[8] For simplicity it is assumed that the minimum efficient scale (MES) for domestic firms is unchanged as industry expansion and contraction alters factor prices, raising and lowering cost curves.[9] This implies that underlying the industry supply curve S in figure 1, all domestic firms are producing at their MES and that movements along this supply curve imply industry entry (movement up and right) and rising factor prices, or exit (down and left) and falling factor prices. The value P^w is the world price at which imports are supplied under perfect elasticity to this small economy. In the initial equilibrium the quantity $Q^D - Q^S$ is imported.

---

7. Maloney and McCormick (1982) show that often technology-based regulation accompanied by entry restrictions produces the same type of cartel-like gains associated with output regulation. And entry restrictions are a natural part of such regulations, usually taking the form of higher, more costly, standards imposed on new entrants.

8. Buchanan and Tullock assumed that the industry was sufficiently small that there were no such factor-price effects, producing a horizontal long-run supply curve. But a horizontal supply curve is not consistent with positive foreign imports. Hence this modification of the Buchanan and Tullock assumptions is a necessary part of the internationalization of their model. Alternatively, an upward sloping long-run market supply curve might be a consequence of firms having different technologies, with all but marginal entrants experiencing positive profits. This case is not considered, however.

9. In general, of course, the scale at which the firm's long-run average cost curve reaches a minimum may increase or decrease as factor prices rise and fall. While this information is not essential to the results, the details of the story are enhanced by working through each of these possibilities. It is well known in the case of technology-based pollution control that such regulation will not necessarily affect all firms identically. Leone (1977) has argued, for example, that the metal finishing industry saw a fivefold increase in the minimum efficient scale due to water quality standards (reported in Oster 1982). Such a change clearly has important implications for market structure, and thus for trade, that are not addressed here.

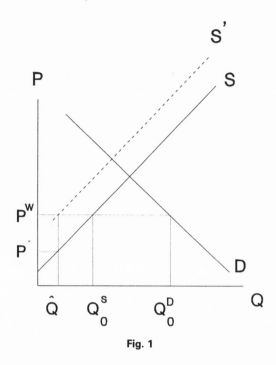

Fig. 1

It is assumed that given existing technologies each unit of output generates a fixed amount of environmental pollution for which no social compensation is paid in the initial equilibrium. Following Buchanan and Tullock (1975) and Maloney and McCormick (1982), we assume that the authorities can accurately measure and monitor these emissions and that they have determined the amount by which they wish to reduce them.[10] Let the desired level of pollution abatement be achieved when domestic production falls to $\hat{Q}$ in figure 1. It is at this point that we assume that lobbying begins. Because the question of how much pollution control has been solved by assumption, the only remaining issue is how to achieve it.[11]

---

10. It should be noted that the ability to measure and monitor emissions falls far short of the information required to determine the optimal level of emissions. For this the authorities would need a well-specified social welfare function which requires detailed knowledge of the preferences of every citizen. So while the amount by which to reduce emissions can easily be set by government agencies, this amount need not approach the social optimum. Once the desired level of emissions has been specified, the best governments can do is to institute policies that achieve that quantity at the least cost.

11. Of course, this is a simplification. Lobbying will necessarily have played a role in determining the desired level of pollution abatement. Our analysis, like that of Buchanan and Tullock, might be construed as investigating the second stage of a two-stage game.

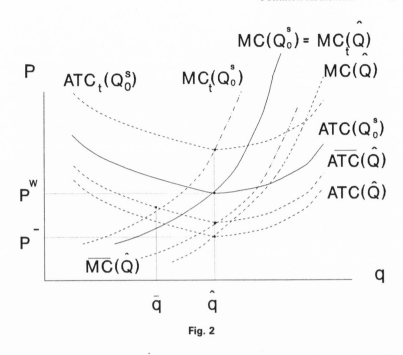

Fig. 2

Consider the imposition of a penalty tax, τ, on units of emissions. The tax is imposed only on domestic firms since the pollution is embodied in production, not in the product itself. This has the initial effect of increasing costs (average and marginal) in the domestic industry by the amount τ. The domestic market supply curve shifts upward by τ to S' in figure 1. Given the constant world price for output, P^w, domestic firms experience transitional losses. In response, exit occurs and total domestic production begins to decline. As this proceeds, factor prices fall and cost curves shift downward. Equilibrium is restored when the factor-price effect has fully offset the effect of the tax on long-run average cost. Firms remaining in the industry are once again achieving normal profits given the net-of-tax price P^-, and producing at the minimum efficient scale. In this sense the pollution abatement is achieved efficiently. The adjustment process at the firm level is shown in figure 2. The initial long-run cost curves are represented by $ATC(Q_0^s)$, and $MC(Q_0^s)$. The argument Q indicates the presence of a factor-price effect. At the initial scale of domestic production these curves shift up by the full amount of the tax to $ATC_\tau(Q_0^s)$ and $MC_\tau(Q_0^s)$. When total production is cut back to $\hat{Q}$, those firms remaining in the industry will have *cum-tax* cost curves corresponding to $ATC(Q_0^s)$ and $MC(Q_0^s)$, costs having shifted down because of the implied decline in factor prices. The new cost curves without the tax appear as $ATC(\hat{Q})$ and $MC(\hat{Q})$.

As an alternative to the penalty tax, policymakers could simply impose production restraints on existing domestic firms. The level $\hat{Q}$ of domestic production could be achieved by assigning equal production quotas to each firm so that $N_0 \cdot \bar{q} = \hat{Q}$, where $\bar{q}$ is the assigned quota and $\hat{Q}$ corresponds to the targeted level of pollution abatement. Such a policy certainly achieves the pollution-abatement objective. It does so, however, inefficiently in the sense that each firm is assigned a production quota below its minimum efficient scale.[12] Inefficiency implies that for a given set of factor prices the regulated industry will require a greater quantity of all factors (if there are no inferior factors) than under the tax-based case. Thus to produce $\hat{Q}$ under the regulatory solution requires a smaller decline in factor demand than is implied under the penalty tax. The factor-price effect, therefore, will cause the ATC curve to shift down, but by less than the amount of a corresponding penalty tax. Another way of expressing this social cost comes from observing that the profit per firm under the output-regulation scheme is necessarily less than the tax revenue per firm under the penalty tax.[13] The social cost of pollution abatement under output regulation, therefore, exceeds that under a penalty tax.

The penalty tax is socially superior to output regulation in another sense.[14] Observe that the equilibrium in the case of regulated output, like any cartel equilibrium, requires enforcement to maintain it. Each firm has an incentive to cheat on its quota allotment, other things equal, since the world price exceeds marginal cost at $\bar{q}$. Hence some monitoring and enforcement expenses must accompany the regulated solution. However, while each firm has an individual incentive to break its quota in the absence of effective enforcement, each also has an incentive to support an effective enforcement mechanism as part of the regulatory package. In addition, regulators must deter entry as part of the regulatory package in order to achieve and maintain the pollution-control objective. Neither entry deterrence nor an enforcement mechanism is required in the case of a penalty tax since normal profitability prevails in the post-tax equilibrium.

Both the penalty tax and output regulation lead to a rise in imports to fill the demand gap at the stable world price. Total imports rise and the domestic industry's market share falls by exactly the same amount under the tax as

---

12. It is noteworthy that had we eliminated the assumption of the constancy of the minimum efficient scale under the factor-price effect, the conclusion that output regulation is inefficient remains, unless by chance the firm's MES declines exactly to the quota level selected by officials. This is highly improbable. Nevertheless, the extent of the inefficiency will depend to some degree on the change in a firm's MES.

13. Looking at figure 2, we see that p^w exceeds $ATC(\hat{Q})$ at $\bar{q}$ by less than the per unit tax. Hence per unit profit is less than per unit tax. Since the same aggregate quantity is produced, $\hat{Q}$, total industry-wide profit under the output regulation must be less than total tax revenue under a penalty-tax scheme.

14. The observations in this paragraph largely follow the observations made in Buchanan and Tullock 1975.

under the quota. However, the demand for sector-specific factors falls by somewhat less under output regulation than in the tax-based case since the aggregate quantity $\hat{Q}$ is produced inefficiently.[15] Sector-specific factor prices also fall by less than in the tax case. In addition, layoffs are distributed evenly across firms under the quantity regulation, whereas all-or-nothing layoffs occur under the penalty tax. Also in contrast to the tax case, there is no industry exit under the output regulation so the market share of each firm must be less than that under the tax. Finally, under quantity regulation each firm has excess capacity, which was not the case under the tax (except during the transition to the new equilibrium). All of these developments offer evidence of injury *due to imports*.[16] The only indication of industry health under output regulation is that profits are up, having gone from normal to supranormal, assuming that the quota does not excessively overshoot the quasi-cartel optimum. In sum, both output regulation and the penalty tax generate layoffs of sector-specific labor (and other factors), sector-specific factor prices fall, market share declines, sales decline, and excess capacity is created (either permanently or in the transition to the new equilibrium) while imports surge to fill the initial demand gap. Indeed, it is clear that the openness of the economy (which prevents domestic prices from rising as they otherwise would) exacerbates the policy-induced decline in the domestic industry.

The presence of injury due to imports (the standards of causality vary from country to country but they are almost uniformly weak)[17] opens a world

---

15. We assume for simplicity that there are no inferior factors of production.

16. In practice, injury is the major necessary condition firms must satisfy in order to obtain import relief. Under U.S. trade law, indicators of injury focus on "health" of an industry as reflected in the levels and trends of production, capacity utilization, market share, inventories, profits, import penetration, price underselling (i.e., the supply price offered by foreign firms is less than that of domestic import-competing firms), and employment. But not all of these indicators need to be affected, and other indicators may also be invoked. Commissioners have a great deal of discretion in deciding which indicators to emphasize and which to downplay (Kaplan 1991). Finger and Murray (1990, p. 39), in looking at the United States unfair trade cases, found that "in almost every unfair trade case that gets to a formal determination, the US government finds that the foreigners are unfair—that the foreign merchandise has been dumped or subsidized. When the US government turns down a petition for an import restriction it is almost always because the injury test is negative."

17. Under Section 201 of the Trade Act of 1974 (the U.S. escape clause), what had once been a fairly demanding standard of causality was weakened substantially. The Trade Expansion Act of 1962 required that an industry prove that it had suffered serious injury, the "major cause" of which was import penetration ("major cause" was defined to mean greater than all other factors combined). Further, the Trade Expansion Act of 1962 required that a specific tariff concession under the General Agreement on Tariffs and Trade (GATT) be cited as causing serious injury. Under the Trade Act of 1974, in order to make relief under the escape clause more accessible, imports need be only a "substantial cause of serious injury, or the threat thereof" (this was interpreted to mean "a cause which is important and not less than any other cause"), and no tariff

of potential government assistance to shelter such firms from foreign competition.[18] One should anticipate some change in the probability of contingent protection under such a scenario, even though this chain of events was set in motion by domestic environmental policy. While evidence of injury caused by import competition is the true prerequisite of protection, not "injury" *per se,* separating injury which is linked directly to foreign competition from that which is attributable only to environmental policy is unlikely to be practicable for trade regulators. In practice, since establishing causality is quite difficult, perhaps impossible, objective evidence of injury is identified without being able to attribute cause with any precision. Thus there is great scope for the "injury" introduced directly by domestic environmental policies to affect the incidence and outcomes of petitions for protection under existing rules.

Interest Group Preferences?

We consider four principal groups with concentrated interests in this area whose voices might be heard in domestic policy circles. These include domestic environmentalists, domestic producers, domestic sector-specific factors (e.g., labor), and foreign producers. Consumers perceive no direct interest since the price of the good is unaffected in the small-country setting. Even when the price is affected, consumer's interests are not sufficiently concentrated for them to be mobilized. Also, in what follows it is assumed that any potential revenues generated from a penalty tax are to become part of a general fund which is to be dispersed evenly across all segments of the economy.

Consider first the case of sector-specific factors. For simplicity of exposition we will treat these factors collectively as a composite of nontradable sector-specific labor. The analytical work above indicates that layoffs will occur under both a penalty tax and output regulation. But because the industry

concession need be identified as causing injury. On this see Jackson (1989, pp. 161–65). Still it must be emphasized that emergency protection is generally a less-preferred path to protection, being dominated by antidumping laws and access to voluntary restraint agreements.

18. Article XIX of the GATT, the so-called *safeguards* provision, offers signatory nations an avenue to escape their GATT obligations and to erect protective barriers to trade under the condition that "product is being imported into the territory of that contracting party in such increased quantities and under such conditions as to cause or threaten *serious injury* to domestic producers in that territory" (Article XIX[1][a]). The action may continue "to the extent and for such time as may be necessary to prevent or remedy the injury" (Article XIX[1][b]). An additional avenue of protection is provided in Article VI of the GATT, supplementary provisions to Article VI, and in the Agreement on Implementation of Article VI of the GATT (commonly known as the Antidumping Code). Again, relief under this article is contingent on a finding of injury; specifically a finding of *material injury* or the threat of such is required. It is generally agreed that material injury is established at a relatively low threshold.

output $\hat{Q}$ is produced inefficiently under the output regulation, industry-wide factor demand falls by less than under the penalty tax. This suggests that while sector-specific factors may oppose the environmental protection in the first place (since factor prices and employment decline), once the decision to intervene has been made, the penalty tax poses a greater threat to employment and wages than does output regulation. Other things equal, this suggests a strict preference for the regulated approach over the penalty tax.

A number of additional factors reinforce this preference. First, the rents accruing to domestic firms under output regulation offer a potential carrot to labor. The prospect of capturing any small share of these rents should again render the quantity restraint the preferred approach. Under the penalty tax no rents accrue to the penalized industry, and so no prospect of an ameliorating outcome for labor is built into the tax scheme. Second, in the case of regulated output, layoffs are distributed evenly across existing firms, while a penalty tax produces all-or-nothing outcomes. That is, under the penalty tax a given firm either lays off its entire work force as it exits the industry or it survives intact with its full contingent of workers undisturbed. If labor and politically aligned groups perceive evenly distributed layoffs to be more equitable (in a self-interested sense) than the all-or-nothing purges of a tax, the strict preference for the regulated approach to pollution control is again strengthened. Since such all-or-nothing layoffs are likely to have regional implications, and since regional labor officials are likely to want to avoid the prospect of their demise, it seems likely that a preference for the evenly distributed layoffs associated with the regulatory approach will emerge over the all-or-nothing option tied to the penalty tax. In addition, the burden of the proportional layoffs under regulation will most likely affect employees with the least seniority. To the extent that the interests of workers with greater seniority are protected over those of more recent vintage, the threat of evenly distributed layoffs under quantity regulation appears attractive relative to the all-or-nothing threat under the penalty tax.

Environmentalists can be expected to value both current and expected future environmental quality, largely to the exclusion of all else. Given such focused preferences, how will environmentalists evaluate a penalty tax versus the quantity regulation? Both achieve the same pollution objective in the static framework. Nevertheless, environmentalists can express reasonable concern with each approach. The penalty-tax approach is attractive since it is self-enforcing. There is no incentive for firms to deviate from the new equilibrium, other things equal. At the same time, there is no reason to be absolutely confident about the stability of the induced reduction in output. After all, firms are not *required* to reduce output and pollution. Their choice remains free but subject to a penalty for each unit of pollution produced. Should there be, for example, an exogenous decline in factor prices, firms can be expected to step up production, and so emissions. The introduction of cost-saving technologies that may be equally polluting will also induce firms to step up production under a fixed penalty tax. If environmentalists weigh the threat of such future developments strongly they may conclude that the flexibility and autonomy remaining in the hands of polluting firms under a penalty tax is undesirable. The penalty tax may not provide sufficient assurance of ongoing pollution abatement to satisfy environmentalists.

The regulatory alternative carries with it the potential problem that it is not self-enforcing. The incentive for each firm to exceed its production allotment is strong in the absence of effective oversight and severe sanctions. This is because at the assigned production quota price exceeds marginal cost for each firm. But this problem is easily overcome. When packaged with a credible enforcement component and effective barriers to entry the regulatory approach is likely to dominate the tax-based abatement scheme from the environmentalist's perspective. It has two principal advantages. First, it provides greater certainty than the tax scheme. The level of pollution abatement cannot rise in this sector because production is controlled directly by government regulators, and violations are penalized severely. The future environmental threat posed by a penalty-tax scheme is eliminated under the regulatory solution. With little distortion of the environmentalist's perspective, the difference might be likened to the option of sentencing a criminal to probation or to prison. In the former case, the criminal's behavior during the probationary period is very likely to be modified so that criminal activity is deterred. In the later case, there is unequivocal restraint that precludes criminal behavior. Environmentalists are likely, therefore, to prefer the binding regulatory solution to the promise of deterrence under the penalty tax.

Another factor pointing to the superiority of the regulatory approach is that environmentalists may be concerned with the cleanup of past environmental degradation. Under the penalty-tax scenario some firms are forced to exit the industry and those that remain earn zero economic profits. Under the quantity regulations no firms exit and each earns supranormal profits. The presence of ongoing supranormal profits provides assurance that these firms will have the wherewithal to correct past abuses. While governments can impose barriers to exit if certain cleanup criteria have not been met, firms still require resources to engage in any cleanup activity. Hence, just as labor might see the rents associated with the regulatory approach as being potentially captured at a later date, so environmentalists may view these rents as assuring latent resources for expected future cleanup.

As emphasized by Buchanan and Tullock (1975), the preferences of polluting firms lie squarely in favor of production quotas with strict enforcement measures and binding barriers to entry since cartel-like profits are potentially available.[19] In practice the barriers to entry selected by governments tend to appease both firms and environmentalists. Frequently, prospective new entrants must meet technology-based pollution control standards that are far more stringent than those imposed on existing firms. The right to use the old technology tends to be grandfathered for existing firms. Potential entrants are thereby placed at a substantial cost disadvantage, making entry all but impossible. As long as the proposed quantity restraint does not overshoot by too much the underlying cartel solution (where firms in the small open economy are able to exercise collective monopsony power in factor markets, not monopoly power), polluting firms will favor such a regulatory package. The environmental objective offers polluting firms a socially attractive pretence for cartelizing their industry. Government graciously agrees to cover enforcement costs and establish barriers to entry to support this enterprise.

The proposition that domestic polluting firms will favor regulation is strengthened when these firms must compete with foreign imports. In the United States, import-competing firms are adept at using the antidumping procedures, Section 301 threats,[20] and rules for emergency protection against foreign rivals to gain strategic advantage. Similarly, import-competing polluters can be expected to identify the strategic opportunities presented by calls for environmental protection. In the present context, the regulatory approach to pollution control becomes additionally appealing to these firms because it enhances the expected present value of the prospect of protection.

It does this first by providing a formal institutional setting for cooperative behavior that reinforces their ability to pursue other (non-pollution-abatement) areas of mutual interest, including protection from foreign competition. As observed in their empirical study of the U.S. steel industry's use of unfair trade laws in the period 1982–86, Herander and Pupp (1991, pp. 143–44) point out that "segments of the industry which have difficulty forming an effective coalition because of the free-rider problem obtain a less favorable policy outcome, allowing for the degree of material injury." Thus on average

---

19. Had the analysis considered a domestic industry that is already highly concentrated and exercising substantial market power, neither form of pollution control may appear attractive. If, however, a monopolized sector saw sufficient value in shifting the ongoing expense of deterring entry to government regulators, the regulatory approach may then be supported in order to capture that prize.

20. Section 301 of the U.S. Trade Act enables domestic firms to allege "unjustifiable or unreasonable" trade actions by foreign governments that impose barriers to market access. If the accusation is supported by the case, special market-sharing arrangements may be negotiated (imposed on) with the foreign government. The European Community has a similar instrument in Regulation 2641/84. See Jackson 1989, pp. 107–9.

it pays to present a united front in petitions for protection. In his analysis of the collusive effects of EC antidumping law, Stegemann (1990, pp. 276–78) also points out the importance that domestic producers "present a common view" in order to enhance their chance of a favorable outcome. Olson (1965, p. 132) observed the importance of such organizational catalysts to rent seeking when he pointed out that "The common characteristic which distinguishes all of the large economic groups with significant lobbying organizations is that these groups are also organized for some *other* purpose." Regulatory oversight under a pollution-control regime offers such import-competing industries this *other* purpose. Internalizing industry-wide incentives, including the incentive to petition for protection, is certainly part of the attractiveness of the regulatory approach.

Second, the regulatory regime establishes a precedent for market sharing that may pave the way to the inclusion of foreign firms. Domestic firms may be able to use the market-sharing arrangement imposed for environmental reasons to argue for its extension to foreign firms. That is, should protection be granted it may be marginally more likely that a negotiated voluntary restraint agreement will be the chosen instrument. Other things equal, such quantity-based protection offers greater opportunities for the consolidation of market power and thus greater profits.

Third, because of the barriers to domestic entry established under the regulatory pollution-abatement scheme, the prospective profits of protection will not be dissipated by competition over time. This means that the regulatory scheme serves to increase the expected present value of the profits of protection, thereby increasing the appeal of protection. No such barriers accompany the penalty tax, and so protection from foreign competition would offer only transitory profits.

Finally, there is the issue of injury. When the above conditions are linked to the induced structural upheaval (layoffs, declining sector-specific factor prices, declining market share, reduced domestic sales, and excess capacity) and the corresponding surge in imports, the probability of a successful petition for protection from this sector under the regulatory approach to pollution control almost certainly rises.[21] In sum, the pressure for protection under the regulatory scheme may be enhanced since the domestic industry experiences

---

21. As mentioned above, the injury criteria established under escape-clause and unfair-trade legislation offer a great deal of discretion in evaluating the health of an industry. See, e.g., Morkre and Kruth 1989 and Jackson 1989. The principle value of injury for VER protection is that it helps to mobilize the political resources needed to induce protective action. That is, once an industry is entitled to protection under existing administrative rules (e.g., unfair trade laws), the prospect of negotiating a VER becomes more likely.

injury in several dimensions, this injury coincides with a surge in imports, regulatory barriers to entry increase the expected capitalized value of protection, and the cooperative behavior enforced by the regulations furthers the industry's ability to speak with one voice.[22]

For those troubled by the proposition that the probability of protection rises under the regulatory approach (by at least as much as under the penalty tax) even though profits are up, a few additional points can be made. In a recent paper, Leidy and Hoekman (1991) argue that because import-competing firms collectively control or influence many of the indicators of injury (employment, sales, capacity utilization, profits, etc.) they are not likely to remain passive in this regard in their pursuit of contingent protection. Managing injury criteria must be regarded as a tool of *indirect* rent seeking. Whether through accounting practices, public bluffs (e.g., announcing the cancelation of a significant capital expansion program), strategic layoffs of workers (sufficient to induce greater political visibility), or the like, firms can influence the political perception of industry health and thereby influence the probability of protection under existing rules. Should such firms perceive that current profits will hurt them in a petition for protection, especially when all other indicators support their case, they will have a strong incentive to adjust their behavior at the margin, giving up some current profits in the expectation of recouping these under future protection. Under the regulatory approach to pollution control a mechanism for overcoming the free-rider problem is already in place. So the industry's ability to act in unison is being facilitated by the regulatory regime. One way that firms might adjust their current profit figures during the protection-seeking stage is for them *voluntarily* to incur costs related to the cleanup of past hazardous dump sites. Such a tactic would serve the dual purpose of winning the ongoing favor of their coalition partners (the environmentalists) while also enhancing their prospect of protection by concealing supranormal profits. So the fact that under passivity industry profits rise need not jeopardize the industry's prospect of protection at all.

Finally, consider the interests of foreign exporting firms. In the small-country case, one might simply assert that "smallness" implies irrelevance from the exporter's perspective. One might also suggest that foreign producers will have less access to the domestic political process than local pressure groups. Neither claim is consistent with recent experience. The United States and the European Community, for example, have consistently taken positions in the domestic policy debates of small countries, particularly when

---

22. Domestic environmentalists are likely to favor such protection, especially if there are international pollution spillovers from the foreign production, since it implies an incidental reduction in foreign pollution.

their trade is affected. And precisely because these trading partners are significant to the small country, the voice of foreign interests tends to be heard.[23]

Recall that price is unchanged and exports rise equally under both the
penalty tax and the regulatory approach, suggesting indifference between the
two regimes at this point. The structural upheaval in the domestic market is
thus only of interest to foreign exporters to the extent that it affects the
probability that barriers to trade will block market access in the future. While
both approaches to pollution abatement generate injury and so enhance the
probability of future barriers to trade, exporter preferences cannot be separated from the type of protection likely to arise. Unlike, for example, tariff
barriers under an act of emergency protection, or quantitative restrictions
where the quota rights are auctioned, VERs are often negotiated so as to
confer rents on both foreign exporters and domestic firms.[24] The prospect of a
VER therefore is often not a threat but an opportunity for foreign exporters to
consolidate market power. Because the market-sharing approach to pollution
abatement increases the likelihood of a mutually advantageous VER being
negotiated, exporting firms should express a strict preference for the regulatory approach.[25] Under the penalty tax, even if it could be argued that a VER
might also be sought and won, the absence of domestic entry barriers still
makes it a less attractive alternative, since any profits that might initially arise
will be dissipated by competition over time. The domestic barriers to entry
combined with the prospect of a mutually advantageous VER, therefore,
make the regulatory solution the preferred approach by foreign exporters.

In the small-country case we conclude that there is likely to be unequivocal support for the inefficient regulatory approach to pollution control from all
of the major (politically influential) interested parties. Much of this support
derives from the likelihood that regulation, with the help of existing rules for
contingent protection, will facilitate the joint cartelization of the regulated
market by foreign and domestic producers. Thus the appropriate measure of
the social cost of the pollution-abatement policy extends substantially beyond
the problem of static inefficiency. Under current rules for contingent protection, pollution control in an import-competing sector may come at the expense
of free trade, unless the policy prescriptions of economists include measures
to preempt these incentives.

---

23. See Hillman and Ursprung 1988 for a recent model of trade-policy formation that
explicitly incorporates foreign interests. Also see the discussion in Hillman 1990.

24. See, e.g., Harris 1985.

25. Hillman and Ursprung 1988 argue that because VERs are a conciliatory trade policy,
whereas tariffs come at the expense of foreign exporters, VERs tend to be chosen over tariffs
whenever that policy choice is available. In the current model the political support for a VER is
strengthened.

2.2. The Large-Country Case

In this section we drop the assumption that domestic policy will have no effect on the terms of trade. All other assumptions are maintained and the same policy options are examined in the large-country setting. Figure 3 describes the initial equilibrium and that corresponding to the introduction of a penalty tax. Figure 3*a* depicts the domestic market. Figure 3*b* captures equilibrium trade. The import demand curve in figure 3*b* (MD) is derived by varying the world price and observing the corresponding excess demand in figure 3*a*. The export supply curve in figure 3*b* (XS) is derived similarly from the market conditions abroad captured in figure 3*c*. Underlying the initial equilibrium (P_0^w, Q_0^w) there are, say, N_0 domestic firms and N_0^* foreign firms, each earning normal profits. The introduction of a penalty tax in the domestic economy is analyzed in figure 3*a* and *b*. There is an upward shift in the domestic market supply curve in figure 3*a* and a corresponding rightward shift in the demand curve for imports indicated in figure 3*b*.

The specific tax has been set to induce the level of pollution abatement implied by the decline in domestic production from Q_0^s to $\hat{Q}^s$. Initially, this implies that the minimum of the long-run ATC for domestic firms rises to the level indicated by point *B* in figure 3*a*. At the initial world price, the home industry experiences losses and exit begins. As exit proceeds and factor demand falls, factor prices decline domestically and the supply side of the economy adjusts from point *B* toward point *C*. Excess demand for imports at price P_0^w causes the world price to rise. This creates transitional supranormal profit abroad which induces entry. As foreign entry proceeds sector-specific factor prices rise abroad. Entry continues until excess profit is eliminated at P_1^w. Observe that unlike the small-country case the product price rises, but by less than the full amount of the tax. The price received net-of-tax is therefore below P_0^w. It should also be clear that in order to attain the same level of pollution control in the large-country case as that for a small country, the penalty tax must be higher due to its effect on world price. Still, firms at home and abroad are once again producing at their MES and earning a normal return. Again, in this sense the pollution abatement is achieved efficiently.

As in the small-country case, the penalty tax induces increased imports while the domestic industry contracts under transitional losses. Lost sales, exit, lost market share, declining sector-specific factor prices, transitional excess capacity, a declining net price, and transitional losses assault the domestic industry following the penalty tax. At the same time, the foreign exporting industry experiences increasing sales abroad, entry, rising sector-specific factor prices, transitional capacity utilization beyond the normal rate, and transitional supranormal profits. This is certainly a recipe for enhanced pressures for protection under existing institutions and legal rules. But as

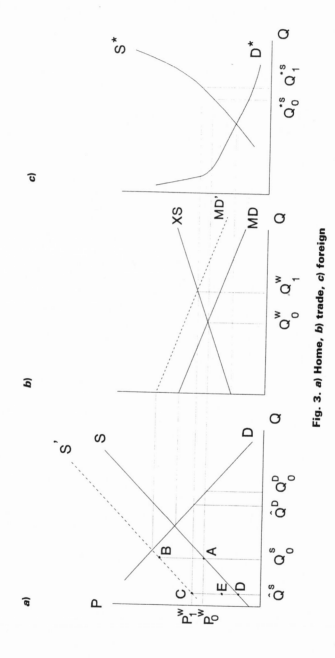

Fig. 3. *a)* Home, *b)* trade, *c)* foreign

in the small-country case we will argue that the efficient pollution tax is likely to be blocked by interest groups.

The effect of allocating production quotas to existing domestic firms can be analyzed in figure 3a. To achieve the level of pollution abatement implied by $\hat{Q}^s$, the regulatory authority assigns the quota $\bar{q}$ to each domestic firm, where $N_0 \cdot \bar{q} = \hat{Q}^s$. This means that the effective market supply curve is given by the original supply curve in figure 3a up to point D, at which point it becomes vertical. The same static equilibrium obtains as under the penalty tax. Aggregate employment and sector-specific wages fall, but by somewhat less than under the penalty tax. Domestic sales decline, market share is lost, and domestic firms will have excess capacity at their assigned production quota. At the same time imports rise, exporters experience unusually high rates of capacity utilization during the transition, transitional excess profits induce entry in the foreign market, and sector-specific factor prices and employment rise while total foreign output expands. Because of the factor price effect, which drives minimum unit costs for domestic firms to a point like E, supranormal profits can be earned if the production regulation is not overly strict.

Since domestic firms are not producing at their MES and because the factor-price effect is less under the quantity regulation, per-unit profit must be less than the value CD, which is the amount of the corresponding penalty tax. The cost conditions of the representative firm appear in figure 4. Prior to the environmental regulation the long-run cost curves are given by $MC(Q_0^s)$ and $ATC(Q_0^s)$. As the output restraint is imposed under the regulatory regime, industry-specific factor prices fall and cost curves shift downward while the market price rises.[26] Any firm-level quota greater than q' yields positive firm profits. But per-unit profit must be less than per-unit tax revenue under the equivalent penalty tax. And since aggregate output is the same under both regimes, total tax revenue must exceed the total profit under the regulatory approach.

Interest Group Preferences?

All of the incentives identified in the small-country case which induce interest groups to support the regulatory approach over the efficient penalty tax continue to hold. But several additional observations can be made. First, to the extent that there are international environmental spillovers, domestic environmentalists may feel that their efforts are being thwarted as foreign production rises. Second, foreign environmental groups, if any, will perceive a beggar-thy-neighbor dimension to the domestic environmental policy as it stimulates

---

26. As distinguished from the small-country case, domestic firms are able collectively to exercise both monopoly and monopsony power under the quantity regulation.

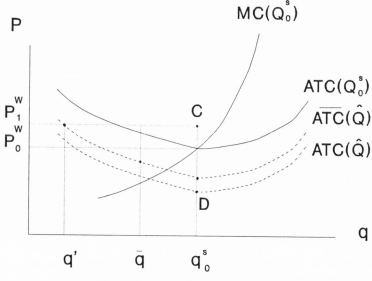

Fig. 4. Representative domestic firm

local polluting activity. Third, foreign firms experiencing transitional supranormal profits may display heightened sensitivity to the problem of free entry. In the transition to the new equilibrium these firms can be expected to resist in political forums the erosion of these newly acquired supranormal profits. And a trade-policy agreement may provide the means to that end.

All of these serve to reinforce support for the regulatory approach to pollution control. Taking the last point first, foreign firms can be expected to be receptive to any policy change (at home or abroad) that increases the likelihood of sustaining the otherwise transitional excess profits associated with their competitor's environmental controls. Under the quantity regulation the increased likelihood of a VER offers foreign firms a prospective vehicle for achieving entry barriers. The existence of VERs is often sufficient to deter entry in a foreign export sector. In other words, what could not be achieved independently through the political process might be achieved with the help of a foreign government intervening on behalf of its import-competing industry to negotiate a VER.

If we assume that foreign environmental groups have been unsuccessful in their direct bids for pollution control, such groups may nevertheless perceive an opportunity to play an indirect role. Both the tax and the regulatory approach induce equivalent increases in foreign production in the static sense. But as argued above, the quantity regulation increases the political pressure for the negotiation of a VER. The prospect of a VER means that the expected level of additional environmental degradation abroad under the regulatory approach will be less than that under the tax-based approach. Indeed, the expected restraint agreement may be sufficiently restrictive as to imply reduced production/pollution abroad. The enhanced prospect of restraining foreign production/pollution through trade policy should make the regulatory approach attractive to both foreign and domestic environmental interests. In this sense, the linkage between the regulatory approach to environmental control and the prospects for a cartelizing VER is part of the attractiveness of the regulatory approach to both firms and environmentalists, at home and abroad.

3. Policy Implications

Throughout the analysis it was assumed that the prospective tax revenues from a penalty tax were to become part of a general fund not specifically designated for any interest group. One obvious way government might shift the balance of lobbying power toward the efficient policy is to credibly precommit a share of these revenues to one or more of the interested parties. It has already been determined that the potential profits under the quantity restraint (without considering the prospective profits of international cartelization) will fall short of the total tax revenues. In principle, it might appear therefore that there is sufficient revenue generated by the tax to "buy" the support of the polluting industry. But this is not the case. If the government imposes a penalty tax and commits to redistributing the revenue in lump sums independent of production (as it must be to maintain efficiency) to polluting firms, free entry will still drive industry profits net-of-tax and net-of-tax-revenue to zero. The appeal of such a scheme falls short of that of the regulatory approach unless it can be combined with entry barriers, e.g., grandfathering the rights to the tax revenues. Hence for polluting firms it is not simply the unattractiveness of the tax scheme that leads them to support the regulatory approach. It is the prospective profits of the regulatory approach that wins their favor. In order to enlist their support for the penalty tax it must cease being a penalty. It must produce an expected net benefit at least as great as the regulatory approach. And this cannot be achieved by any possible redistribution of tax revenues alone.

Environmentalists, though, might be persuaded to shift their support under a credible plan to divert a share of tax revenues to the environment. Recall the original source of environmentalists' support for the regulatory approach. Principally, it was argued that the regulatory approach provides greater certainty of a favorable environmental outcome. The tax-based approach leaves it to firms to decide how much to cut back. If a share of total tax revenues were set aside for environmental causes, any autonomous increase in industry production will generate new revenues for environmentalists at the same time it increases direct pollution. What was perceived as a fault before is now a mixed blessing. Hence such a plan may undermine the primary source of environmentalist opposition to the penalty tax approach.[27]

There is, of course, still no guarantee that the interests of domestic import-competing firms, not to mention foreign exporters and sector-specific labor groups, will be blocked under such a policy. In fact, the expected economic profit under the regulatory approach provides firms with the resources needed to bribe environmentalists for their support in the same way that government might use tax revenue to that end. And the increase in the probability and expected profitability of protection under the regulatory approach supplements the industry's ability to secure environmentalist support. So the proposal that a share of any environmentally based tax revenues be committed to environmental causes is probably necessary but not sufficient to render the efficient policy politically feasible. Finally, it must be emphasized that any scheme designed to enlist support for the "efficient" approach to environmental tasks, because it will inevitably require a diversion of resources to that end, implies social costs that must be weighed against the social costs attached to the alternative inefficient policy.

Ultimately one must ask whether the social cost of doing nothing is sufficiently great to warrant the attention of policymakers. In addition to the static losses due to cartelization, because of the probabilistic linkage of environmental regulation to contingent protection, the full social cost of "doing nothing" must include the differential effect of the trade tax versus the regulatory approach on both the probability and the expected type of protection. We have argued that not only is the probability of winning protection higher under the regulatory solution, but this approach might produce the more pernicious and less transparent form of protection, VERs. But the actual social opportunity cost of clinging to the status quo is not the difference between social welfare under the efficient versus the inefficient policies. A Pareto improvement over the status quo will be achieved if a new policy formulation can guarantee the political feasibility of the efficient tax without imposing social costs of its own in excess of those associated with the quantity regulation.

The process of change itself is costly, as it necessarily involves the costs of lobbying to influence change, the costs attached to resisting such influence, the cost of resisting change altogether, and ultimately the diversion of re-

---

27. In a related vein, Reuter's (March 1991) recently reported that the European Environmental Bureau (EEB), a federation of 130 environmental groups, expressed their support for a European energy tax. The EEB's proposal, however, sets aside one-third of the tax revenues for spending on environmental policies at the EC and national levels.

sources to those activities intended to secure the prescribed change. Since these costs are appropriately treated as inherent to any policy change, the social opportunity cost of abiding the flaws of the status quo must be less than a comparison of welfare under any two idealized states would suggest. Simply put, if the cost of the institutional change needed to render the efficient environmental policy politically feasible exceeds the loss experienced under the inefficient policy, then no institutional change is desirable. A certain resistance to institutional/legal change is, therefore, well grounded in optimizing principles. For change to be warranted, current inefficiencies will have to be relatively great to justify incurring the costs (and uncertainties) inherent in change.

With this caveat in mind, there are, of course, other ways to tame rent seekers who block efficient policies.[28] Since inefficiency implies forgone income, one approach may be to inform the general public of that loss. While the benefits to pressure groups under an existing inefficient system are concentrated and the social costs dispersed, there can be no hope of overcoming the free-rider problem unless the costs to consumers and/or the direct benefits to special interests are made well known. An example reported originally in *Fortune* and discussed in Koford and Colander (1984, 207) suggests that in some cases a low-cost information campaign may be enough to change the political balance. Apparently the leading voice in a lobbying operation to prohibit the use of polyvinyl chloride (PVC) pipes through revised building codes was a steel company with a recently diminished market share in the conduit market. Its share had fallen from 50 percent to 32 percent due principally to competition from PVC pipes. While the lobbying strategy emphasized the company's concern over the purported fire hazard of PVC pipes, public knowledge of this conflict of interest would at least create a political atmosphere in which such moral posturing would be exposed and deflated.

Policy recommendations might therefore focus on mobilizing prospective countervailing forces that might favor efficient solutions to environmental externalities. Provision of objective information is likely to be a crucial necessary condition to achieving this. There is regrettably no reason to expect that the increased pressure for protection induced by environmental policies will be countervailed endogenously. Indeed, one might argue that antiprotectionist forces[29] are likely to be weakened as they may want to avoid being labeled "anti-green" by environmentalists.

---

28. Koford and Colander 1984 suggest a menu of possibilities.

29. See, e.g., Destler and Odell 1987 for a discussion of the political economy of anti-protection.

4. Concluding Comments

The connection between environmental policy and trade policy arises in the first instance directly out of the injury criteria for protection. But we also argued that the inefficient environmental regime strengthens the trade-policy linkage in several other ways. It sets a precedent for market sharing that may be extended to foreign firms, making a VER marginally more likely. It establishes barriers to entry that will preclude the dissipation of the profits of protection, thereby increasing its appeal and inducing more petitions for protection, other things equal. And it also provides an institutional framework that may assist in the presentation of a unified front when seeking protection. The deadweight costs of inefficient environmental policies applied in import-competing sectors, therefore, may be compounded by the social costs of the administered protection they help to induce.

Experience with national environmental policies, indeed regulatory policy in general, suggests that what *should* be done on Pareto efficiency grounds may not be politically feasible. What *will* be done need not be welfare superior. The political economy of the regulatory process teaches us that it is not enough to formulate "optimal" or efficient policies in the abstract. Policy relevance requires that such solutions be packaged with those institutional-legal changes needed to make them politically feasible. If the lessons of this model are robust, there is likely to be a strong and pervasive distaste for efficiency among interest groups in the area of environmental policy. And this aversion appears to be exacerbated by existing institutions and practices in commercial policy, particularly the rules for contingent protection. Interest-group support for the inefficient policy is not terribly surprising since efficiency implies the absence of rents, which are the lifeblood of such groups. The challenge for policymakers (and policy economists) is to redirect the efforts of pressure groups toward socially desirable policies without also initiating new opportunities for rent seeking.

REFERENCES

Banks, Gary. 1990. "Australia's Antidumping Experience." World Bank Working Paper WPS 551, December.
Buchanan, J. M., and Tullock, G. 1975. "Polluters' Profits and Political Response: Direct Controls versus Taxes." *American Economic Review* 65 (March): 139–47.
Destler, I. M., and Odell, J. S. 1987. *Anti-Protection: Changing Forces in United States Trade Politics.* Washington, D.C.: Institute for International Economics.
Finger, J. M., and Murray, T. 1990. "Policing Unfair Imports: The United States Example." *Journal of World Trade* 24:39–55.

General Agreement on Tariffs and Trade. 1989. "Developments in the Trading System." September 1988–February 1989.

General Agreement on Tariffs and Trade. 1991. *Trade Policy Review: The European Community,* vol. 1, June.

General Agreement on Tariffs and Trade. 1990. *Trade Policy Review: The United States of America,* March.

Harris, R. 1985. "Why Voluntary Export Restraints Are 'Voluntary.'" *Canadian Journal of Economics* 18:799–809.

Herander, M. G., and Pupp, R. L. 1991. "Firm Participation in Steel Industry Lobbying." *Economic Inquiry* 29:134–47.

Hillman, Arye L. 1990. "Protectionist Policies as the Regulation of International Industry." *Public Choice* 67:101–10.

Hillman, A. L., and Ursprung, H. W. 1988. "Domestic Politics, Foreign Interests, and International Trade Policy." *American Economic Review* 78, no. 4:29–45.

Jackson, J. 1989. *The World Trading System.* Cambridge, Mass.: MIT Press.

Kaplan, S. 1991. "Injury and Causation in USITC Antidumping Determinations: Five Recent Views." In Matthew Tharakan, ed., *Policy Implications of Antidumping Measures.* Amsterdam: North Holland.

Koford, K. J., and Colander, D. C. 1984. "Taming the Rent-Seeker." In David C. Colander, ed., *Neoclassical Political Economy.* Cambridge, Mass.: Ballinger Publishing Co.

Leidy, M. P., and Hoekman, B. M. 1991. "Spurious Injury as Indirect Rent Seeking: Free Trade under the Prospect of Protection." *Economics and Politics* 3, no. 1 (July): 111–37.

Leidy, M. P., and Hoekman, B. M. 1993. "'Cleaning Up' While Cleaning Up? Pollution Abatement, Interest Groups, and Contingent Trade Policies." *Public Choice* 78:241–58.

Leone, R. 1977. "The Real Cost of Regulation." *Harvard Business Review,* November–December, 57–66.

Low, P. 1992. "Trade Measures and Environmental Quality: The Implications for Mexico's Exports." In Low, P., ed., *International Trade and the Environment,* World Bank Discussion Papers 159, pp. 105–20. Washington: World Bank.

Maloney, M. T., and McCormick, R. E. 1982. "A Positive Theory of Environmental Quality Regulation." *Journal of Law and Economics* 25 (April): 99–123.

Morkre, M. E., and Kruth, H. E. 1989. "Determining Whether Dumped or Subsidized Imports Injure Domestic Industries: International Trade Commission Approach." *Contemporary Policy Issues* 7, no. 3:78–95.

Olson, M. 1965. *The Logic of Collective Action.* Harvard Economic Studies, Harvard University Press.

Oster, S. 1982. "The Strategic Use of Regulatory Investment by Industry Sub-Groups." *Economic Inquiry* 20:604–18.

Stegemann, K. 1990. "EC Anti-Dumping Policy: Are Price Undertakings a Legal Substitute for Illegal Price Fixing." *Weltwirtschaftliches Archiv* 126, no. 2:268–98.

Stigler, G. J. 1971. "The Theory of Economic Regulation." *Bell Journal of Economics and Management Science* 3, reprinted in *The Citizen and the State: Essays on Regulation,* University of Chicago Press, 1975.

Tobey, J. A. 1990. "The Effects of Domestic Environmental Policies on Patterns of World Trade: An Empirical Test." *Kyklos* 43, no. 2:191–209.

United States International Trade Commission. 1989. "Gray Portland Cement and Cement Clinker from Mexico." USITC Publication no. 2235, November.

Yandle, B. 1989. *The Political Limits of Environmental Regulation.* New York: Quorum Books.

CHAPTER 4

Jobs versus Wilderness Areas: The Role of Campaign Contributions

Dennis Coates

The belief that money has corrupted the political process, especially with regard to the making of public policy, is widespread. Consequently, campaign finance reform proposals have proliferated in the last twenty-five years. Systematic research into the effect of campaign contributions on the voting positions of members of Congress has addressed the statistical significance of contributions variables in determining the vote of a legislator. The research has not taken the next step, however, which is to ask whether a change in contributions of a given size would (*a*) alter the outcome of the vote, or (*b*) change the "position" taken on the broadly defined issue by a given legislator. If the answer to both questions is "no," then the effects of money on the policy process are much less worrisome. On the other hand, if the answer to either question is "yes," then this chapter provides support for those who want to reform campaign finance.

Our results indicate that the answer to the first question is clearly no, at least for the policy examined here. Simulations of the impact of a change in the level of contributions reveal a very weak sensitivity to contributions. For example, only 18 representatives changed their votes on one of the bills studied after receipt of an addition $3,300, the sample mean, from timber interests. One might suggest that buying 18 votes for less than $60,000 is a bargain. However, in this instance, 18 votes is not sufficient to alter the outcome. Moreover, the buyer must know which 18 votes are for sale at this price. In the simulations, all legislators are given the additional contributions, resulting in a total expenditure for the purchase of those 18 votes of over a million dollars. Looked at in this way, these votes are not, perhaps, so cheap. Additional contributions from corporate PACs of $31,100, also the sample

This chapter was improved thanks to the comments of participants in the Institute for Research in Social Science Political Economy Working Group at the University of North Carolina, Garey Durden, and, especially, Roger Congleton. The aforementioned individuals are, of course, blameless for any remaining errors.

mean, influenced the votes of only 32 legislators, again not enough to alter the final outcome. Results of other simulations are generally of equal or smaller magnitude. The implication is, therefore, that public policy, in this instance, is not noticeably influenced by campaign contributions.

Concerning the ability of contributions to alter the "position" of the recipient, the results indicate an answer of "yes, but." Probit estimates of both influences on individual votes and on an index of "ideology" find that campaign contributions are statistically significant, providing the "yes" portion of the answer. On the other hand, contributions are found to be correlated with the unobserved influences on a legislator's "ideological position"; that is, contributions are given both to influence the legislator and to assist those legislators whose "ideology" is similar to that of the contributor. This latter effect explains the "but" portion of the answer.

Taken together, the results of this analysis suggest that campaign contributions do affect legislators' positions and their voting patterns. The problem, however, is not such that the money actually sways public policy, at least at the voting stage. It is possible that contributions influence the content of bills, the degree of aggressiveness with which legislators and their staff lobby for the contributors, and in other ways not readily made quantitative. This research does not address these latter issues.

Our analysis focuses on the voting in the House of Representatives on amendments to wilderness designation legislation for federal lands in California and Oregon. Wilderness designation means, under the 1964 Wilderness Act, that federally owned lands will be protected from commercial development, petroleum and mineral leasing, logging and road building, and off-road vehicle recreation. Representative Robert Walker, R-Pa., proposed the amendments to the designation bills to allow the secretary of agriculture to waive wilderness designations shown to raise unemployment.[1] The debate could be framed, therefore, in the jobs versus the environment language familiar from the 1992 presidential campaign. The analysis here focuses on the role that campaign contributions play in influencing the voting on these amendments, and on the selection of "jobs versus wilderness areas" ideology by legislators.

The analysis utilizes two interesting insights from the recent literature. The first of these is attributable to Thomas Stratmann (1991). Building upon the work of Chappell (1982), Stratmann argued that contributors' motives may be either to help (re)elect a candidate whose views are akin to their own, or to influence the policy position of a candidate whose opinion differs from

---

1. Douglas Booth (1991) examines the role that dependency of a region on the timber industry plays in the allocation of federal roadless lands to wilderness, nonwilderness, and future planning status by both the Forest Service and Congress. He concludes that Congress is more sensitive to this dependency than is the Forest Service; Congress allocated relatively fewer areas to wilderness status in timber-dependent Oregon, but relatively more in not-dependent Washington, than were recommended by the Forest Service.

theirs. He suggests that one may infer which motive prevails from the sign of the correlation between the unobservable influences on contributions received and votes cast. If the correlation is positive, then contributions are motivated by the desire to enhance the likelihood of election of a desirable candidate; if negative, contributions are an attempt to sway the opinion of a legislator.

The second insight is provided by Durden, Shogren, and Silberman (1991). They construct an index which measures the position of a legislator on the issue of strip mining regulation. Specifically, they form a discrete representation of a continuous issue space. They are able to order their representation by determining precisely which bills a legislator voted for and against. Their approach is quite different from that of Kalt and Zupan (1984) or Pashigian (1985), each of whom simply recorded whether a vote was or was not in agreement with the position of some group or individual. The problem with the Kalt and Zupan and Pashigian approach is that a legislator may vote against a bill favored by a special interest group because, in that legislator's estimation, it is too "liberal" or too "conservative." Simply aggregating all the "for" votes and all the "against" votes need not accurately characterize the "ideology" of the voter.

To better see this point, consider the two amendments which are the subject of this chapter. The Oregon amendment allowed the U.S. Department of Agriculture to waive, at its own discretion, wilderness provisions that could be shown to raise unemployment. However, for the agriculture department to waive wilderness designations in California, the state must first have requested such action. Therefore, a legislator with one yes and one no vote may be indicating quite different preferences depending on which amendment was voted for and which against. Following the Durden et al. approach, these two amendments allow construction of a discrete ranking of the intensity of pro-jobs/antienvironmental feeling on the part of House members. Legislators could vote no on both amendments, yes on both, no on the California amendment but yes on the Oregon amendment, or yes on the California and no on the Oregon amendment. These four possible voting records are ranked according to their inherent "jobs versus environment" sentiment. This ranking is then analyzed using an ordered probit technique.

The study is organized as follows. The history of wilderness area policy is reviewed in section 1. Section 2 describes the theoretical model(s) of legislative decision making that inform the empirical work. Section 3 describes the data and the estimation scheme. Section 4 presents the results, and section 5 concludes.

1. United States Wilderness Policy

The Second Roadless Area Review and Evaluation (RARE II) program inventoried 46.1 million acres of undeveloped national forest lands in the lower 48 states. The purpose of the evaluation was to decide which areas should be preserved as wilderness. The evaluations began in 1977, and President Carter made recommendations to Congress in 1979. His recommendations were that 9.9 million acres in 36 states be designated wilderness and 28.5 million as nonwilderness. An additional 7.7 million acres were set aside as further planning areas.

Before Carter left office, Congress had acted on his recommendations for seven states, designating 4.2 million acres as wilderness. The 97th Congress continued the work of responding to the RARE II and Carter recommendations setting aside an additional 132,000 acres in six states. In February of 1983, the Reagan administration decided to scrap the RARE II evaluations and begin anew, prompted largely by a federal court ruling (*California v. Block*) that the environmental impact statement (EIS) on which the recommendations were made was inadequate. Environmentalists and their allies in Congress settled on allowing the wilderness designation bill of each state to include a clause declaring "sufficient" the EIS on which the recommendations were based.

Controversy surrounded the Oregon designation because the Oregon congressional delegation was divided, along party lines, on its appropriateness. Republican representatives Dennis Smith and Robert Smith opposed designation of the area as wilderness. Dennis Smith offered an amendment which, if passed, would have designated no area in Oregon as wilderness, exempted the Oregon EIS from judicial review, and released all nonwilderness lands in Oregon for other uses. When that failed (58 to 292), Representative Don Young, R-Alaska, proposed an amendment that would exclude any land in the district of either of the Smiths from designation as wilderness. This amendment also failed (91 to 249). Rep. Walker offered an amendment that would have allowed the secretary of agriculture to waive any provisions of the bill shown to raise unemployment. Walker's proposal was defeated with 96 favorable and 240 unfavorable ballots. Finally, the bill designating 1.13 million acres in Oregon as federal wilderness came to a vote. The vote was 252 in favor to 93 opposed.

When Congress took up the issue of designating wilderness areas in California, two factors impinged on the voting. The first of these was the fact that the bill designated as wilderness in excess of a million acres more than had been recommended by the Forest Service and the Reagan administration. The principal proponent of the larger amount was Representative Phillip Burton, D-Calif. His death two days before the voting is thought to have swayed votes toward accepting the larger acreage. There were attempts to reduce the acreage being set aside and to weaken the protections just as there had been prior to the voting on the Oregon designation. Specifically, Representative Walker again offered an amendment that would have allowed the secretary of agriculture to waive provisions of the bill shown to increase unemployment. Unlike his amendment in the Oregon case, however, such action by the secretary would only be possible if the state of California requested it. The outcome was not changed; the amendment was defeated 121 to 272. Representative Norman Shumway, R-Calif., proposed amending the designation to adhere to the recommendation of the Reagan administration. The Shumway proposal was defeated 136 to 257.

2. Models of Legislator Decision Making

2.1. Choice of Roll Call Vote

The analysis of legislator voting decisions has been carried out to date primarily on the basis of a verbal description of a utility maximization model (Stigler 1971; Kau and Rubin 1979, 1981; Kalt and Zupan 1984, 1990). The difficulty with this approach is that it makes it easy to omit important influences. For example, the focus in each of the mentioned studies, and many more, is on the costs and benefits of the relevant piece of legislation to the legislator's district. This style of analysis assumes that elected representatives are empowered only with the ability to signal their constituents' desires. The finding by Kau and Rubin (1977, 1978) and Kalt and Zupan (1984) that something other than constituent characteristics and districtwide costs and benefits influence the voting of representatives clearly shows that this delegate model of representation is inadequate.

A better model of representation is of a trustee, whose job it is to make up his or her own mind about the best course of action, and vote accordingly. In such circumstances, the choices made reflect the preferences, or tastes, of the representative, as well as his or her electoral circumstances and the preferences of the constituency. In other words, the legislator's voting decision is the result of a constrained optimization. However, few empirical models of representatives' voting decisions account for either the preferences or the electoral circumstances of the legislator. In this study, these factors are explicit.

For a single roll call vote the legislator weighs the costs and benefits of a vote for the bill and of a vote against the bill. The theoretical depiction of this process is given by the indirect utility function whose arguments are the districtwide costs and benefits, the legislator preferences, and the electoral circumstances of the legislator. Represent the (empirical) indirect utility function as $I^* = f(D, P, E, C) + \epsilon$, where I^* is an unobservable index of utility, $D, P, E,$ and C are vectors of district costs and benefits, legislator preferences, electoral circumstances, and campaign contributions, respectively, and ϵ is an unobservable effect on utility. The legislator votes in favor of the legislation if

I^* exceeds zero, but against the bill otherwise. What is observed is either a vote for the bill $I = 1$, or a vote against the bill $I = 0$. Therefore, the empirical model is

$$I = 1 \Leftrightarrow f(D, P, E, C) + \epsilon > 0,$$

$$I = 0 \Leftrightarrow f(D, P, E, C) + \epsilon \leq 0.$$

(1)

This model is estimated for each of the Walker amendments.[2]

Chappell (1982), Stratmann (1991), and others point out that campaign contributions C may be correlated with the error term in this equation.[3] For example, those unobservable influences that induce a legislator to vote favorably on a bill may also induce larger (or smaller) contributions from PACs favoring support of that bill. If such is the case, estimates of the parameters defining the function $f(.)$ will be biased and inconsistent. Chappell (1982) and Stratmann (1991) estimate simultaneously a tobit equation for contributions and a probit voting equation. Interestingly, Stratmann finds consistent support for the influence of contributions on votes, whereas Chappell (1982) finds the opposite. On the other hand, Stratmann finds that correlation exists between the errors in the contributions and voting equations in four out of the ten cases, twice positive and twice negative. Chappell, however, finds significant positive correlation in five out of eight models. In no case does Chappell find a negative and significant correlation.

The correlation between the unobservables across equations takes on importance beyond controlling for the endogeneity of contributions in the voting equation in Stratmann's analysis. He argues that the sign of the correlation between contributions and voting behavior indicates the motivation of the contributor. For example, if the objective is to influence the election outcome, then a positive correlation will exist between unobserved influences on the contributions and unobserved influences on voting. In this case, the contributor is attempting to help elect someone whose preferences are similar to his or her own. If, on the other hand, the intention is to influence the "position" or, in Stratmann's terminology, platform, of the legislator, then the error in the contributions equation will be negatively correlated with the error in the voting equation. He concludes from his results on the signs and significances of the correlations, therefore, that in some cases contributors attempted to ensure election of their preferred candidate and in other cases attempted to buy votes.

---

2. The functional form of the $f(.)$ function is made explicit below.

3. Becky Morton and Charles Cameron (1992) provide a nice review of the theoretical campaign contributions literature.

Stratmann's hypothesis about the correlation is an interesting attempt to address the issue of contributor motivation. It is likely, however, that inference of motivation based on this correlation is unfounded for two reasons. First, legislators receive contributions from many types of organizations, often with competing interests. For example, the legislation studied in this chapter, amendments to wilderness designation bills whose intent is to protect jobs, has clearly identifiable opponents and proponents. Representatives are likely to receive contributions from both groups; there are, therefore, two (or more) potentially endogenous right-hand-side variables. Both Chappell and Stratmann restrict their attention to issues in which benefits of the legislation are highly concentrated but costs are diffuse. In this way, they are able to avoid the complication of competing interest groups.[4] With multiple contributors the control for endogeneity of contributions becomes more complicated. Nonetheless, endogeneity must be accounted for if the parameter estimates are to be consistent. Moreover, if there are omitted variables, such as contributions from other groups in the Chappell and Stratmann analyses, then the coefficient estimates, including the correlation, are biased.

The second reason the correlation may not be used to infer contributor motivation is related to the manner in which contributions enter the voting equation. Using contributions only linearly, as Chappell (1982) and Stratmann (1991) have done, imposes the restriction that an additional dollar of contributions from a given group has the same impact whether that group has already given \$100 or \$100,000. Such an assumption should be tested, especially since it violates the usual assumption of diminishing marginal utility (productivity). Moreover, if contributions are endogenous and enter the equation in higher order form, then the higher order variables are also endogenous. Hence, the correlation between the error in the contributions equation and the error in the voting equation is a function of the manner in which contributions enter the voting equation. The analysis that follows tests both for the endogeneity and nonlinear impact of contributions.

2.2. Choice of Platform or Ideology

Much has been written in economics and political science about politicians' choice of position in an issue or ideological space. Durden, Shogren, and Silberman (1991) provide an interesting approach to inferring this choice from the observed votes of a legislator. The votes made by a politician reflect the multitude of factors which are modeled in equation 1 above. On any n bills the number of possible voting records is, if abstentions are not possible, 2^n. Each of these 2^n voting records is a revealed ideological position. If n is small

---

4. Chappell explicitly recognizes this shortcoming.

enough, and the subject matter of the bills sufficiently similar, then it is possible to order these voting records from right to left, or conservative to liberal, or "less of activity X" to "more of activity X."[5] In this case, a given legislator's votes provide a discrete representation of his or her position in that space.[6]

In the current instance, the ideological space is a line the extremes of which indicate whether the legislator will choose, under all circumstances, "jobs over wilderness" or "wilderness over jobs." The line, therefore, represents "positions" one could take on the question of tradeoffs between saving/creating jobs or preserving wilderness areas.[7] In what follows, a discrete representation of this space is constructed from legislator votes on the Walker amendments offered to the Oregon and California wilderness designation bills. Recall that these amendments differed slightly. In particular, the California amendment allows the secretary of agriculture to waive wilderness designations only if such a waiver has been requested by the state. The Oregon amendment contains no such restriction. The precise construction of the index is discussed below. For the moment assume that an ordered index exists.

Let I_i^* be the unobservable position of legislator i along this ideological continuum. Of course, I_i^* is unobservable. If there are four possible outcomes, as there are in this case in which $n = 2$, then the model is

$$I^* \equiv f(D, P, E, C) + \epsilon < \mu_0 \Rightarrow I = 0,$$

$$\mu_0 \leq I^* \equiv f(D, P, E, C) + \epsilon < \mu_1 \Rightarrow I = 1,$$

$$\mu_1 \leq I^* \equiv f(D, P, E, C) + \epsilon \leq \mu_2 \Rightarrow I = 2,$$

$$\mu_2 \leq I^* \equiv f(D, P, E, C) + \epsilon \Rightarrow I = 3.$$

(2)

After omitting the I^* and rearranging, the probability that I takes on each of the discrete values can be determined. These probabilities are

---

5. Hinich and Pollard (1981) and Enelow and Hinich (1984) describe a technique by which positions in a high dimension issue space may be reduced to positions in a single dimensional "ideology" space.

6. An alternative method of characterizing the position in the issue space of a legislator is the approach of Kalt and Zupan (1984). Their dependent variable is the log of the ratio between the number of votes favoring regulation and the number opposed. The advantage of the approach used here is that the exact nature of the bill is relevant to characterizing the position of the individual. That is much less true under Kalt and Zupan's method. Their approach has the advantage of easily handling a large number of votes; that is, their dependent variable is continuous.

7. Implicitly, preserving wilderness areas puts jobs, especially in extractive industries, at risk. The analysis does not admit of the possibility that preserving wilderness areas creates jobs.

$$Prob(I = 0) \equiv Prob(\epsilon < \mu_0 - f(D, P, E, C)),$$

$$Prob(I = 1) \equiv Prob(\epsilon < \mu_1 - f(D, P, E, C))$$

$$- Prob(\epsilon < \mu_0 - f(D, P, E, C)),$$

$$Prob(I = 2) \equiv Prob(\epsilon < \mu_2 - f(D, P, E, C))$$

$$- Prob(\epsilon < \mu_1 - f(D, P, E, C)),$$

$$Prob(I = 3) \equiv Prob(\mu_2 - f(D, P, E, C) < \epsilon).$$

(3)

where the i subscript has been dropped for simplicity. Assuming that ϵ is normally distributed with mean zero and unit variance allows estimation of the parameters of this model, including the μ's, using ordered probit.

This model enables the researcher to examine the impact of campaign contributions on the (revealed) position of the recipient in the issue space. If the critics of the role of PAC money are correct, then one should find that money matters but little else does. On the other extreme, PAC money may be completely irrelevant to the preferred policy position of the legislator. Finally, it may be that PAC funds influence the legislator's thinking at the margin, moving the legislator slightly up or down in the continuum. In this latter case, a further interesting question arises. To wit, are the marginal impacts sufficient to push the legislator from one to another of the discrete indicators? Alternatively, for a given increase in receipts, how many legislators become discernibly more (less) pro-jobs or pro-wilderness? This is, of course, the interesting policy question. The movement to reform campaign finance and conventional wisdom take it as a given that campaign contributions influence public policy.

3. Data and Estimation Scheme

Before describing the data, it is important to point out that the lack of individually significant coefficients with prespecified signs is not an indictment of the analysis for at least two reasons. First, the equations estimated are in reduced form. This means, of course, that inferences on the direction of influence of the variables is very difficult. For every story suggesting that a coefficient be positive there may be an equally plausible story implying it be negative. In other words, hypotheses on the signs of coefficients are not necessarily meaningful. To make this point concrete, the discussion of the variables which follows offers alternative rationales on their direction of influence. Second, the emphasis is on the impact of campaign contributions on the platform and

TABLE 1. Descriptive Statistics

Variable and Abbreviation	Oregon Voting	California Voting	Index
South SOUTH	.27 (.45)	.26 (.44)	.27 (.44)
West WEST	.22 (.42)	.23 (.42)	.23 (.42)
Age AGE	49.7 (11.0)	49.3 (10.7)	49.6 (11.0)
Male MALE	.95 (.22)	.95 (.22)	.95 (.22)
Lawyer LAWYER	.49 (.50)	.49 (.50)	.50 (.50)
Democrat DEM	.65 (.48)	.59 (.49)	.64 (.48)
Vote share VOTEPC	68.0 (13.8)	67.2 (13.7)	67.7 (13.8)
Terms in office TERMS	3.7 (3.8)	3.6 (3.7)	3.7 (3.9)
Democratic presidential vote share DEMPRSPC	42.12 (11.19)	41.27 (10.78)	41.68 (11.01)
Median age MEDAGE	30.1 (3.5)	30.1 (3.4)	30.1 (3.5)
Median house value HSEVAL	50.79 (21.43)	51.10 (22.10)	51.13 (21.32)
Median income MEDINC	20.65 (12.27)	20.67 (11.40)	20.75 (12.59)
Percent blue collar PCTBLU	31.5 (7.1)	31.5 (7.2)	31.5 (7.1)
Percent urban PCTURB	74.0 (22.6)	73.9 (25.8)	73.8 (22.6)
Environmental contributions ($) ENVIRO	365 (999)	338 (977)	366 (1,015)
Timber contributions ($) TIMBER	3,317 (3,828)	3,547 (4,072)	3,373 (3,864)
Corporate contributions ($) CORP	30,447 (27,124)	32,278 (29,922)	31,065 (27,457)
Labor contributions ($) LABOR	324 (1,104)	325 (1,180)	331 (1,133)
Observations	340	400	321

voting record of the recipient. To get unbiased estimates of this impact, the model most not omit important influences on the legislator's voting decision. Inclusion of irrelevant variables, while leading to inefficiency, does not bias the estimates. Moreover, it is easy to test for inclusion of irrelevant variables. Many such tests are performed and reported.

The data for this analysis comes from three sources. First, characteristics of the legislator and his or her district were taken from the *Almanac of American Politics*. The roll call votes on the two amendments come from the *Congressional Quarterly Almanac: 1983*. Finally, the campaign contributions data are taken from the Inter-University Consortium on Political and Social Research (ICPSR) tapes of the Federal Election Commission records for 1981 and 1982.

Tables 1 and 2 provide variable definitions and descriptive statistics for the independent variables in the analysis. The explanatory variables fall into three basic types. First, there are district characteristics which serve as proxies for the interests of the constituency. These variables are *SOUTH, WEST,*

TABLE 2. Descriptive Statistics on Contributions

	Oregon Voting	California Voting	Index
Environmental	365 (999)	338 (977)	366 (1,015)
contributions	[1,363; 1,544; 91]	[1,367; 1,569; 99]	[1,382; 1,582; 85]
Timber contributions	3,317 (3,828)	3,547 (4,072)	3,373 (3,864)
	[3,889; 3,867; 290]	[4,112; 4,111; 345]	[3,952; 3,900; 274]
Corporate contributions	30,447 (27,124)	32,278 (29,922)	31,065 (27,457)
	[30,810; 27,079; 336]	[32,604; 29,895; 396]	[31,457; 27,406; 317]
Labor contributions	324 (1,104)	325 (1,180)	331 (1,133)
	[1,103; 1,819; 100]	[1,181; 2,018; 110]	[1,144; 1,879; 93]
Observations	340	400	321

MEDAGE, PCTBLU, PCTURB, MEDINC, and *HSEVAL. SOUTH* and *WEST* are dummy variables indicating that the legislator represents a district from a Southern or Western state, respectively. Intuition might suggest that the coefficients on these regressors be positive in the probit equations for the two amendments, and negative in the ordered probit model because representatives from these regions are more conservative or probusiness than representatives from other regions. However, the signs may be reversed because the population of the south and west is more outdoors oriented than that of the north and east. Therefore, the signs of the coefficients on *SOUTH* and *WEST* are theoretically ambiguous.

The median age in the district is captured by *MEDAGE.* Older populations may be more pro-jobs than others because they suffered from the economic upheaval of the Great Depression, suggesting a positive sign on this variable in the amendment equations but a negative sign in the ideology model. Alternatively, older populations may be more pro-wilderness than younger populations as they reflect on the cleanliness of the environment in the days of their youth. The proportion of the district's workforce employed in a blue-collar job is represented by *PCTBLU.* Blue-collar workers might be expected to be more concerned with jobs than with wilderness preservation; therefore *PCTBLU* could have a positive sign in the single vote equations but a negative sign in the "position" model. On the other hand, blue-collar workers may be more likely than white-collar workers to hunt or fish in the wilderness areas, signaling a desire on their part to protect the environment.

PCTURB measures the percentage of the population living in urban areas. Pashigian's (1985) hypothesis that legislators from urban areas vote so as to impose competitive disadvantages on rural areas does not seem plausible in this case. Jobs lost in the timber and mining industries will not relocate to urban districts. However, contraction of the timber and mining industries could cost jobs in manufacturing located in urban areas. It is plausible, then, that greater urbanization will lead to greater pro-jobs sentiment. On the other hand, urban residents may favor protection of wilderness areas for recreation and "option-value" reasons. If this is true, greater urbanization may imply greater support for wilderness preservation. Therefore, the sign on *PCTURB* is ambiguous.

If wilderness preservation is a normal good, then higher income districts will be more pro-wilderness. Under this hypothesis, an increase in *MEDINC*, which measures the median income of a district, will lead to stronger pro-wilderness sentiment. *HSEVAL*, the median house value in the district, measures wealth effects. On the other hand, housing values are likely to proxy for intangibles such as community stability, ethnic or racial composition, political activism, and others. The sign on this variable is, therefore, ambiguous.

A second type of explanatory variable is the legislator characteristic. The literature on legislator voting typically uses some measure of the individual's voting record to control for ideology. This type of control is, however, weak for several reasons (Jackson and Kingdon 1992). Coates and Munger (1995) argue that legislator characteristics are a reduced form proxy for "ideology." Therefore, the legislator's *AGE*, and dummy variables indicating whether the representative is *MALE*, or a *LAWYER*, capture the role of preferences. The legislator's share of the vote in the previous election, *VOTEPC*, and his or her years of service *TERM* represent the political constraints on the behavior of the representative. *VOTEPC* indicates the electoral security of the member of Congress, for example. The greater is this security, perhaps, the freer to vote his or her own conscience the legislator will feel. On the other hand, a high vote share might indicate that the legislator has been a faithful agent and is, therefore, likely to vote the district's wishes. The number of terms of service also captures some of this security, for it is clear that long-time members are rarely defeated, but also might proxy for institutional pressures. Long years of service might indicate stronger ties to the party hierarchy, or greater likelihood to vote the party line, for instance. Two additional variables capture these security and institutional influences but might also proxy for constituents' interests or ideology. These variables are *DEMPRSPC* and *DEM*. *DEMPRSPC* is the share of the presidential vote received by Jimmy Carter in 1980. *DEM* is a dummy variable indicating a Democrat, if equal to one, or a Republican if zero. Both of these variables are expected to be positively correlated with pro-wilderness sentiment. Therefore, they should have negative signs in the amendment equations but positive signs in the ideology model.

Table 2 presents descriptive statistics for the last type of explanatory variable in the analysis, campaign contributions. Means are reported on these variables for the full sample and for those who receive position contributions. Four types of contributors are distinguished: environmental groups, PACs associated with timber and paper products firms, all corporate PACs, and PACs affiliated with labor groups.[8] The mean contribution received by a legislator in the sample, followed by its standard deviation, makes up the first line in each cell of table 2. The numbers in square brackets on the second line of a cell are the mean and standard deviation for only those legislators who received contributions from that type organization. The third number in the brackets is the number of such recipients.

Table 2 clearly shows that the mean contributions across votes are very similar both in the full samples and in the recipients-only samples. More importantly, the table also reveals that timber and corporate contributions are received by the vast majority of legislators; therefore, the means across the full and recipients-only samples are very similar. The recipients-only sub-samples for labor and environmental contributions are, on the other hand, much smaller than in the full sample. Typically, in this data set, only about 25 to 35 percent of the legislators receive contributions from labor and environmental groups. The upshot of this relatively small number of recipients is that the means across samples vary widely; generally, the recipients-only sample mean is about three times that of the full sample. Mean contributions among recipients-only for labor and environmental groups remain well below those for timber and corporate interests, however.

Intuition suggests that environmental groups contributions should raise the level of pro-environment sentiment of the recipient, and corporate and timber contributions should reduce it. The effect of labor contributions is not so clear, however. Labor groups traditionally give predominantly to Democrats. Hence, it is possible that labor contributions will be positively correlated with the pro-wilderness sentiment of legislators. On the other hand, the interests of labor are to protect and enhance timber and lumber jobs. Therefore, labor contributions may reduce pro-wilderness feelings of legislators.

The dependent variable for the ordered probit model is a discrete representation of the continuous issue space measuring the degree of pro-jobs/antienvironment feeling. Table 3 displays the ranking and the number of legislators falling into each category. Intense pro-wilderness/anti-jobs sentiment is revealed by two negative votes; a person voting this way will not accept discretionary exemptions of the wilderness designations to save jobs. Two affirmative votes on the Walker amendments are evidence of a moderate pro-jobs/anti-wilderness position. A negative vote on the Oregon amendment and a positive vote on the California proposal is evidence of a moderate pro-wilderness/anti-jobs stance.

The most extreme anti-wilderness/pro-jobs position is evidenced by an affirmative vote on the Oregon amendment and a negative vote on the California amendment. This pair of votes signals someone who cannot accept any

---

8. The Appendix lists the PACs whose contributions comprise the group totals.

TABLE 3. Voting on Unemployment Waiver Amendments

	Oregon	California	Observed (percent)
Anti-wilderness/pro-jobs 0	Yes	No	18 (5.6)
1	Yes	Yes	77 (24.0)
2	No	Yes	12 (3.7)
Pro-wilderness/anti-jobs 3	No	No	214 (66.7)
Total			321

restrictions on the ability of the federal government to save/create jobs by circumventing the wilderness designations. One might question why anyone who voted for the Oregon amendment, only to see it fail, would vote against the similar, though weaker, amendment in California. One possibility is that the model does not account for the full dimensionality of the "space." For example, this position may signal a degree of anti-state's-rights sentiment, as it removes any role for the state within whose boundaries the wilderness area is found. Alternatively, Roger Congleton suggested that such voting may be predominantly found among the California delegation. In this case, this pattern may be indicative of California politics or district interests. However, only four of the eighteen legislators whose voting falls in this category are from California. None are from Oregon, and fourteen are Republicans. Moreover, because the Oregon amendment was defeated easily, the no vote in the California case may be a cheap signal of one's extreme pro-jobs sentiment. In any event, there is no obvious connection among these individuals to suggest any influence other than unwillingness to weaken the ability of the federal government to protect jobs.[9]

The functional form of $f(D, P, E, C)$, used in either the role call or index equations estimated below is:

$$f(D, P, E, C) = D\beta + P\pi + E\delta + C\lambda + C^2\gamma \tag{4}$$

where β, π, δ, λ, and γ are vectors of parameters to be estimated.[10] The model allows the influence of campaign contributions to vary depending on the size of the contribution. The idea is that the first dollar of contributions is

---

9. Regression analysis was performed under two additional approaches. First, the 18 individuals with this voting record were included in the next most pro-jobs group. Second, these individuals were omitted altogether. Neither of these alternatives changes the fundamental conclusions described below.

10. Alternatively, the estimated equation uses the natural log of the contributions as a regressor. This has the advantage of reducing the number of regressors, especially important because contributions are thought to be endogenous, at the cost of imposing opposite signs on, and a precise relationship between, the first and second order effects.

likely to have a larger impact than the last. Hence, for a given contributor (environmentalists, timber interests, corporations, or labor groups) the corresponding elements of λ and γ should have opposite sign. Moreover, in the roll call votes on the Walker amendments, a rise in the indirect utility indicates a decrease in pro-wilderness/anti-jobs sentiment; that is, it raises the probability of a vote in favor of the amendment. In these circumstances, the λ corresponding to environmental groups should be negative and the γ positive. For timber and all corporate contributions the expectation is that the λ's are positive but the γ negative. As described above, the impact of labor contributions is theoretically ambiguous. The hypotheses are reversed in the case of the index model. Pro-environmental sentiment rises as the index increases. Therefore, the λ's and γ's in the index model should have signs opposite to those in the roll call models.

In this model, there are eight potentially endogenous variables to contend with because there are four contributing groups whose donations enter linearly and are raised to the second power. The following analysis utilizes Rivers and Vuong's (1988) (RV) two-stage conditional maximum likelihood estimator to consistently estimate the probit model with endogenous explanatory variables. RV conduct Monte Carlo experiments which suggest that in small samples their approach performs better than the alternative limited-information methods. In addition, their method is extremely simple to implement.

The RV method first estimates reduced form equations for the endogenous explanatory variables, the errors from which are used as regressors in the probit equation. A simple test of exogeneity is whether the fitted errors from the first stage equation are statistically significant in the second stage. If they are, then the null hypothesis of exogeneity is rejected. The coefficients on the fitted errors are estimates of the correlation between the reduced form error and the probit error.

Regressors in the first stage equations include all of the exogenous variables in the probit equation and interactions among these variables. The interactions are sufficient in number to identify each of the eight endogenous variables. This approach to identification is arbitrary, of course, as it depends upon the ad hoc exclusion of the interaction variables from the probit equations. Nonetheless, given the nature of the data available for capturing the interests of the legislator and his or her constituency, and the political constraints on the legislator, any exclusion restrictions will be arbitrary.

4. Results

4.1. Estimations

The results of probit estimation of the individual roll call votes are presented in table 4. (Tables 5 and 6 contain the results of restricted models on the

TABLE 4. Probit Results

	Oregon Voting	California Voting
CONSTANT	4.49 (2.9)	−.238 (2.2)
SOUTH	.343 (.39)	.685 (.29)*
WEST	.745 (.51)	.448 (.40)
AGE	.003 (.02)	.029 (.01)*
MALE	−.993 (.81)	1.05 (.57)**
LAWYER	−.521 (.27)*	−.603 (.21)*
DEM	−2.29 (.41)*	−2.43 (.35)*
VOTEPC	.009 (.01)	.015 (.01)**
TERMS	−.003 (.05)	−.091 (.04)*
DEMPRSPC	−.041 (.02)**	−.039 (.02)*
MEDAGE	−.013 (.04)	.010 (.03)
HSEVAL	−.002 (.01)	−.015 (.01)
MEDINC	−.097 (.06)	−.005 (.05)
PCTBLU	.010 (.03)	.039 (.022)**
PCTURB	−.010 (.01)	−.0001 (.01)
ENVIRO*10	−.065 (.04)	−.014 (.01)
ENVIRO SQUARED*10^{-5}	.106 (.09)	.021 (.02)
TIMBER*10^2	−.024 (.01)**	−.012 (.01)
TIMBER SQUARED*10^{-4}	.009 (.008)	.004 (.01)
CORP*10^3	.069 (.027)*	.036 (.01)*
CORP SQUARED*10^{-2}	−.005 (.002)**	−.002 (.0001)*
LABOR*10^3	−31.4 (3300)	−.195 (.42)
LABOR SQUARED*10^{-5}	.302 (32.0)	.005 (.005)
LOG LIKELIHOOD	68.03	104.07

*Variable significant at the 5% level.
**Variable significant at the 10% level.

Oregon and California votes, respectively.) The results reveal that campaign contributions are important influences on how the legislator votes. A likelihood ratio test rejects the null hypothesis of joint insignificance. Moreover, endogeneity of the contributions variables is rejected for voting on both the Oregon and California amendments. Note also that for each contributor group the linear and squared contributions terms have opposite sign. This means that the marginal influence of an additional dollar of contributions approaches zero as contributions rise. Moreover, because contributions enter in quadratic form, the total influence of contributions by some type donor may eventually decline. That is, if contributions rise too high, the recipient may be influenced to do the opposite of what the contributor intended. Such results occur in the simulations reported below suggesting that some donors are close to their influence maximizing level of contributions to some recipients.

Additionally, individual legislator characteristics and district sociodemo-

TABLE 5. Oregon Probit Results

	No Contributions	Contributions and Party
CONSTANT	4.17 (2.4)**	−.29 (.34)
SOUTH	.83 (.33)*	
WEST	.63 (.42)	
AGE*100	.43 (1.5)	
MALE	−.39 (.60)	
LAWYER	−.46 (.23)*	
DEM	−2.75 (.33)*	−1.91 (.25)*
VOTEPC*10	.18 (.09)*	
TERMS*10	−.64 (4.4)	
DEMPRSPC*10	−.47 (.19)*	
MEDAGE	−.03 (.04)	
HSEVAL*100	−.15 (1.1)	
MEDINC	−.08 (.05)	
PCTBLU*100	.73 (2.4)	
PCTURB*100	−.73 (.81)	
ENVIRO*10		−.56 (.31)**
ENVIRO SQUARED*10^{-5}		.89 (.14)
TIMBER*100		−.13 (.11)
TIMBER SQUARED*10^{-4}		.42 (.54)
CORP*1000		.65 (.20)*
CORP SQUARED*10^{-2}		−.50 (.18)*
LABOR*1000		−.37 (3.4)
LABOR SQUARED*10^{-5}		.36 (33)
LOG LIKELIHOOD	81.75	86.02

*Variable significant at the 5% level.
**Variable significant at the 10% level.

graphic variables are important. Party affiliation has the strongest individual impact of any variable. However, in both equations whether or not one is a lawyer, and Jimmy Carter's vote share in 1980, have statistically significant effects. Interestingly, many personal and district characteristics are individually significant in voting on the Walker amendment to the California designation bill but are not significant influences on voting on the Oregon bill. In either instance, however, the variables are jointly significant.

Coefficients on individual variables, recall, are frequently theoretically ambiguous. *SOUTH* and *WEST* each have positive signs in the roll call equations, suggesting legislators from these areas are more pro-jobs than their colleagues. *PCTBLU* is also consistent with this hypothesis. The sign of *MEDAGE* is negative in the Oregon case but positive in the California vote. In neither instance is it even remotely accurately estimated; t-statistics are about .3 in absolute value. *MEDINC* has negative sign in both equations, consistent

TABLE 6.   California Probit Results

	No Contributions	Contributions and Party
CONSTANT	−1.14 (1.9)	−.004 (.23)
SOUTH	.89 (.27)*	
WEST	.46 (.34)	
AGE*10	.25 (.12)*	
MALE	.98 (.51)**	
LAWYER	−.46 (.20)*	
DEM	−2.52 (.27)*	−1.90 (.22)*
VOTEPC*10	.14 (.08)**	
TERMS*10	−.97 (.41)*	
DEMPRSPC*10	−.42 (.16)*	
MEDAGE*100	−.03 (32)	
HSEVAL*10	−.14 (.09)	
MEDINC*100	−.19 (34)	
PCTBLU*10	.37 (.20)**	
PCTURB*100	.07 (.49)	
ENVIRO*100		−.22 (.11)*
ENVIRO SQUARED*10^{-6}		.33 (.22)
TIMBER*100		−.90 (.79)
TIMBER SQUARED*10^{-4}		.22 (.40)
CORP*1000		.30 (.10)*
CORP SQUARED*10^{-2}		−.13 (.06)*
LABOR*100		−.88 (.52)**
LABOR SQUARED*10^{-5}		.12 (.06)*
LOG LIKELIHOOD	114.94	129.84

*Variable significant at the 5% level.
**Variable significant at the 10% level.

with the hypothesis that wilderness preservation is a normal good. It is not significant in either case, however. The sign on *HSEVAL* was theoretically ambiguous but in both equations is negative. In the California case, *HSEVAL* has a t-statistic of 1.3. *PCTURB* is negative but insignificant in both roll call equations. The negative sign, however, suggests that urban legislators vote to protect the recreational use of wilderness lands for their constituents.

Table 7 presents the results of the ordered probit estimation. (Table 8 presents estimation results from the restricted models.) The first column contains the results without correction for the endogeneity of contributions. In this case, only the party affiliation variable is significant at the 5 percent level. Its sign is positive, indicating that Democrats are much more pro-wilderness/ anti-jobs than Republicans. This result comes as no surprise. The squared value of labor contributions is significant at the 10 percent level. More importantly, the contributions variables are jointly significant. The district and

TABLE 7. Ordered Probit Results[a]

	Model 1	Model 2	
CONSTANT	−.807 (2.7)	2.03 (14.8)	
SOUTH	−.170 (.35)	5.96 (3.90)	
WEST	.035 (.48)	3.09 (3.47)	
AGE*100	−.507 (1.66)	.830 (11.6)	
MALE	.633 (.44)	.575 (1.5)	
LAWYER	.222 (.25)	.791 (1.4)	
DEM	2.21 (.39)*	−6.64 (7.7)	
VOTEPC*100	−.925 (.84)	2.75 (11.7)	
TERMS*100	−.492 (4.3)	8.83 (35.0)	
DEMPRSPC	.013 (.02)	.009 (.16)	
MEDAGE*10	.099 (.50)	−.161 (.206)	
HSEVAL	−.013 (.01)	−.106 (.10)	
MEDINC	.095 (.06)	.499 (.54)	
PCTBLU*100	.141 (2.76)	−.544 (8.0)	
PCTURB*100	.956 (1.05)	4.76 (11.7)	
ENVIRO*100	.420 (.35)	−.346 (1.4)	.786 (1.3)
ENVIRO SQUARED*10^{-5}	−.064 (.24)	.269 (.55)	−.336 (.46)
TIMBER*1000	.059 (.13)	4.31 (2.8)	−4.28 (2.82)
TIMBER SQUARED	−1.34 (7.03)	−31.07 (14.7)*	31.06 (14.8)*
CORP*1000	−.036 (.03)	−.630 (.44)	.596 (.44)
CORP SQUARED	.264 (.22)	3.55 (2.9)	−3.30 (2.8)
LABOR*1000	.789 (.54)	11.05 (21.1)	−10.2 (21.3)
LABOR SQUARED*10^{-5}	−.014 (.01)**	−.117 (.19)	.103 (.19)
μ_1	1.82 (.19)*	1.95 (.22)*	
μ_2	2.11 (.21)*	2.26 (.23)*	
LOG LIKELIHOOD	159.54	149.77	

[a] μ_0 is normalized to zero in the estimation.
*Variable significant at the 5% level.
**Variable significant at the 10% level.

legislator characteristics, other than party affiliation, appear to add very little. The largest *t*-statistic of any of them is 1.56 with most falling below one. Nonetheless, the restriction that these coefficients simultaneously equal zero is easily rejected at the 5 percent level.

As in the roll call equations, the signs of coefficients are frequently theoretically ambiguous. Representatives from the south are less pro-wilderness (more pro-jobs) than others, but those from the west are more. Districts with older populations are more, not less, pro-wilderness as are districts with large blue-collar populations. In neither case, however, is the *t*-statistic bigger than .5, so these coefficients are extremely imprecisely estimated. *MEDINC* has the expected positive sign, suggesting again that wilderness preservation is a normal good.

TABLE 8. Ordered Probit Results[a]

	No Contributions	Contributions and Party
CONSTANT	−.77 (2.2)	1.71 (.36)*
SOUTH	−.35 (.27)	
WEST*10	.02 (4.2)	
AGE*10	−.11 (.14)	
MALE	.33 (.37)	
LAWYER	.21 (.21)	
DEM	2.50 (.35)*	1.92 (.20)*
VOTEPC*10	−.14 (.07)**	
TERMS*100	.78 (4.1)	
DEMPRSPC*10	.18 (.21)	
MEDAGE*10	.19 (.44)	
HSEVAL*10	−.10 (.10)	
MEDINC*10	.76 (.57)	
PCTBLU*100	−.11 (2.1)	
PCTURB*10	.10 (.09)	
ENVIRO*100		.45 (.37)
ENVIRO SQUARED*10^{-6}		−.72 (2.7)
TIMBER*1000		.17 (.93)
TIMBER SQUARED		.13 (.54)
CORP*1000		−.36 (.20)**
CORP SQUARED		.31 (.17)**
LABOR*1000		1.13 (.39)*
LABOR SQUARED*10^{-6}		−.17 (.07)*
μ_1	1.74 (.17)*	1.68 (.15)*
μ_2	1.99 (.18)*	1.93 (.16)*
LOG LIKELIHOOD	171.46	171.95

[a]μ_0 is normalized to zero in the estimation.
*Variable significant at the 5% level.
**Variable significant at the 10% level.

As before, highly urbanized districts are more pro-wilderness. Pashigian (1985) found that legislators from urban and northeastern districts were decidedly more likely to vote for the policy of prevention of significant deterioration (PSD) of the air quality.[11] He concluded that such votes were evidence of an attempt by legislators from these areas to reduce the relative attractiveness to businesses of high-air-quality, low-tax areas in the western and southern states. The results here show that legislators from more urbanized districts are more pro-wilderness than others. Pashigian's explanation does not work well in this case because the logging and mining displaced or disallowed from the wilderness areas would not relocate to the urban districts. Moreover, to the

---

11. Pashigian's dependent variable is the log of the ratio of favorable to unfavorable votes.

extent that lumber and minerals become more expensive, urban districts will suffer higher consumer prices or loss of jobs in industries using those materials. The vote seems more likely to be motivated by the high option and recreation values placed on wilderness areas by people from urban districts.[12]

The campaign contributions variables as a group are jointly significant. Moreover, in each case the linear and quadratic terms have opposite sign. In addition, a test of the restriction that the quadratic terms are jointly insignificant easily rejects the null hypothesis. The prediction that the first dollar of contributions has a larger impact than the last, therefore, is borne out by the data. Previous work by Chappell (1982) and Stratmann (1991) is, therefore, probably misspecified, casting doubt on their results. On the other hand, the signs on timber contributions are opposite those that theory would predict. That is, one would expect contributions from lumber and paper firms to reduce the extent of pro-wilderness sentiment, but these results show the opposite. The t-statistics on these variables are, however, quite small.

Recall that contributions may be correlated with the error in the probit equations. All of what has gone before may be suspect due to the bias imparted by this correlation. The second column of Table 7 shows the results from an ordered probit equation estimated using the RV consistent method.[13] First, one cannot reject the endogeneity of the contributions variables. Second, the estimated values of the μ's are very similar to those from the simple ordered probit equation. However, the other coefficient values are quite different between the endogeneity corrected equation (model 2) and the uncorrected equation (model 1).

Many of the coefficient estimates in model 2 are, however, implausible. Consider, for example, the implication of the value of the coefficient on *SOUTH* 5.96. This value means that an individual from the south has a value of the index I^* that is 5.96 larger than an otherwise identical individual not

---

12. Interestingly, the option value argument may also apply in Pashigian's case. Representatives from low-air-quality areas may be voting to impose PSD on high-air-quality areas because their constituents value highly the option of visiting or retiring to such areas. His comparison shows that representatives of southern and western districts vote differently on PSD policy relative to automobile emissions standards, but those from urban and northern districts do not.

13. A reduced form equation was estimated for each of the eight contributions variables. Interestingly, there does not appear to be a general pattern of influences on contributions. In other words, those factors which influence contributions from one group need not play a significant role in the other equations. Party affiliation, vote share, the number of years of service to Congress, age and gender of the representative, the district's electoral support of Jimmy Carter in 1980, its level of urbanization, and the region of the country one represents are significant in one equation or another. Among the higher order terms used to identify the fitted values the squared values of vote share, house value, urbanization, and the interaction of party affiliation and Democratic presidential vote share are each individually significant in at least one equation. R squareds range from about .10 to .28.

from the south. This value virtually guarantees that a southerner will be in the most pro-wilderness category, whose cutoff value is 2.26. A similar argument applies to the (wrong-signed) coefficient on Democrat. The value of -6.64 virtually guarantees that a Democrat will be in the least pro-wilderness category, the cutoff value for which is zero.

A likely explanation for the poor results from endogeneity corrected model 2 is the small sample size. Guilkey, Mroz, and Taylor (1992) found that the small sample properties of all the limited-information simultaneous probit estimators are quite weak. One gives up a great deal of accuracy in the coefficient estimates, they conclude, when one uses a consistent estimator, such as the RV approach, rather than a simple probit technique. In their Monte Carlo work, they found the simple probit estimator nearer the true value than the maximum likelihood estimator in more than 60 percent of the replications under some models. For this reason, the simulations reported below use the coefficient estimates from model 1.

Before turning to a discussion of the policy simulations, consider Stratmann's (1991) suggestion that the sign of the correlation ρ between the unobservable influence on voting and the unobservable influence on contributions indicates the motivation of the contributor. In this model, the unobservable influence on contributions enters the voting equation in two places, not in one as in Stratmann's analysis. However, looking only at the fitted errors on the linear terms, one sees that the sign is positive for environmental and corporate contributions and negative for labor and timber contributions. These results suggest, following Stratmann's argument, that environmental and corporate contributors generally hope to help their preferred candidates win election. Labor and timber interest contributors, on the other hand, hope to influence elected officials to vote their way. Note, however, that the coefficient on the quadratic error, in every case, is of the opposite sign from the linear term. This suggests that the direction of the correlation between the error in the contributions equation and the error in the "position" equation is ambiguous. In fact, if contributions enter the "position" equation in log form, the coefficients on the fitted errors are negative for environmental and timber contributions but positive for corporate and labor contributions. In other words, the implicit motivation of environmental and labor groups is opposite what the linear terms from model 2 imply.

Hence, the coefficient estimates on the fitted residuals in model 2 are not estimates of ρ. Instead, an estimate of the correlation between the unobservables is a function of the coefficients on the linear and quadratic error terms. The important point here is not whether ρ is positive or negative, but rather that inference of the goals of contributors from the sign of the correlation between the errors is highly suspect.

Finally, table 9 illustrates the predictive power of the ordered probit

TABLE 9.   **Predictive Power of the Ordered Probit Equations**

Category	Actual	Predicted: Model 1 (Model 2)			
		1	2	3	4
1	18	0 (4)	16 (12)	0 (0)	2 (2)
2	77	2 (2)	67 (65)	0 (0)	8 (10)
3	12	0 (0)	6 (6)	0 (0)	6 (6)
4	214	0 (0)	14 (10)	0 (0)	200 (204)

models. The first model correctly predicts the position of 267 of the 321 legislators. All of the correct predictions come from the two most populous classifications, however. Model 2 correctly predicts the position of 273 of the 321 individuals. Moreover, correct predictions occur in three of the four categories. However, not too much should be made of this increase in the number of correct predictions because model 2 includes eight additional regressors over model 1.

4.2. Simulations

The empirical results reported above suggest that contributions influence the voting decisions of legislators on specific bills. In this instance, however, voting does not influence the extent to which contributions are received. On the other hand, contributions were found to influence and be influenced by the legislators' position in the issue space. None of this evidence, however, speaks to the important policy question of whether contributions influence observed public policy.[14] To address the impact of contributions on policy outcomes several simulations were performed. For each legislator, contributions from a given type PAC were increased both by $100 and by the average contribution, and the policy position of the legislator was determined. For example, after increasing each legislator's receipts from environmental groups by $100, I^* was calculated, call this value I^{**}. If the value of I^{**} fell between minus infinity and μ_0, the legislator was assigned a predicted position of category 0. If I^{**} fell between μ_0 and μ_1 the legislator was predicted to be in category 1, and so on.

Table 10 presents the number of individuals whose vote, or revealed position, changes for a given increment to an explanatory variable. For the

---

14. The voting on each of the Walker amendments was quite lopsided against passage. It may be that legislators were able to cast symbolic or informational votes at little cost because the outcomes were both easily and accurately predicted. In this case, one would expect contributions to have little impact on the voting.

**TABLE 10. Legislators Whose Vote or Position Changes
with Changes in Independent Variables**

	Change in Independent Variable	Oregon Voting	California Voting	Index
Environmental	100	20	11	16
	366	88	36	96
Timber	100	3	0	0
	3,372	9 (9)	18 (17)	8 (8)
Corporate	100	0	1	0
	31,100	30 (3)	32	12 (4)
Labor	100	96	0	5
	331	98 (97)	4	13 (12)
House Value	1,000	0	0	0
	10,000	4	14	5
Median Income	1,000	5	0	7
	4,000	13	0	14
Percent Urban	10%	5	0	7
Vote Share	10%	4 (2)	9	3

Note: Figures in parentheses are number whose position moves opposite the expectation.

most part, the simulations suggest that few votes could be bought and few platforms altered by increased campaign contributions. If each legislator receives a $100 increment to contributions from environmental groups, for example, only 20 legislators vote differently on the Oregon bill, 11 on the California bill, and 16 reveal a different "position" in the wilderness/jobs issue space. If the $100 comes from corporate contributors, only one legislator votes differently, in the California case, and none reveal a different position. Timber contributions of $100 purchase three votes on the Oregon amendment but have no impact on the voting on the California bill or the legislator's platform. A $100 increment in the contribution of the labor groups raises the pro-wilderness voting record of only five individuals. However, that $100 purchases 96 votes on the Oregon bill, but none on the California bill.

One hundred dollars is about one-third the mean for environmental and labor contributions, but only about 3 percent of the mean timber contribution and .3 percent of the mean corporate contribution. Therefore, a second simulation was run in which each legislator received an increment to contributions from a given type contributor equal to the average contribution by that type. For example, in addition to actual receipts from corporate contributors, each legislator was given another $31,100 of corporate contributions. The results of these simulations is further evidence of the insensitivity of legislators' policy positions to campaign contributions. In the case of corporate PACs, only eight individuals become less pro-wilderness, while four become more pro-wilderness. (More about this below.) Timber contributions raise the pro-wilderness positions of eight representatives but reduce no one's. Increasing labor contributions by the mean raises the pro-wilderness position of 12 representatives. If environmental groups increase their contributions to each legislator by the mean the impact on pro-wilderness positions is large. Ninety-six individuals' positions rise, 94 of them by two full categories; that is, someone who was extremely (moderately) anti-wilderness/pro-jobs is influenced to become moderately (extremely) pro-wilderness/anti-jobs.

The anomolous decrease in pro-jobs sentiment resulting from an increase in corporate contributions remains to be explained. Note that the quadratic form of the effect of contributions of a given type allows for the marginal impact to change from negative to positive or positive to negative. For such a shift to occur after a given increase in contributions, the recipient must be "near" the turning point, that is, the level of contributions where the marginal impact reaches zero. The level of corporate contributions at which the marginal impact changes from negative to positive is about $69,231. For the simulated increase in contributions (an additional $31,100 of corporate funding) to reduce the influence of corporate financing, the recipient must receive more than $53,681 before the donation. In the "ideology" sample, 63 individuals have initial receipts above this amount; 32 have receipts above $69,231, that is, beyond the peak level of influence before the simulation.

5. Conclusion

We have examined the influence of campaign contributions on the voting record of a legislator and upon his or her "ideological position." The research has taken a step not previously found in the literature on the effect of campaign contributions, the simulation of the effects of additional contributions on the voting behavior of recipients. These simulations enable one to address the questions of whether the contributions alter either the vote or the "ideology" of the legislator.

The results indicate that the answer to the first question is clearly no, at least for the policy examined here. Simulations of the impact of a change in the level of contributions reveal a very weak sensitivity to contributions. For example, only 18 representatives changed their votes on the California amendment after receipt of an additional $3,300, the sample mean, from timber interests. One might suggest that buying 18 votes for less than $60,000 is a bargain. However, in this instance, 18 votes is not sufficient to alter the outcome. Moreover, the buyer must know which 18 votes are for sale at this price. In the simulations, all legislators are given the additional contributions, resulting in a total "purchase price" for the 18 votes of over a million dollars. Looked at in this way, these votes are not, perhaps, so cheap. Additional contributions from corporate PACs of $31,100, also the sample mean, influenced the votes of only 32 legislators, again not enough to alter the final outcome. Results of other simulations are generally of equal or smaller magnitude. The implication is, therefore, that public policy, in this instance, is not noticeably influenced by campaign contributions.

Concerning the ability of contributions to alter the "position" of the recipient, the results indicate an answer of "yes, but." Probit estimates of both influences on individual votes and on an index of "ideology" find that campaign contributions are statistically significant, providing the "yes" portion of the answer. On the other hand, contributions are found to be correlated with the unobserved influences on a legislator's "ideological position"; that is, contributions are given both to influence the legislator and to assist those legislators whose "ideology" is similar to that of the contributor. Moreover, simulations indicate that the ideology of fewer than 15 representatives is altered by the contributions by timber and corporate PACs. These latter two results explain the "but" portion of the answer.

Taken together, the results of this analysis suggest that campaign contributions do affect legislators' positions and their voting patterns. The problem, however, is not such that the money actually sways public policy, at least at the voting stage. It is possible that contributions influence the content of bills, the degree of aggressiveness with which legislators and their staff lobby for the contributors, and in other ways not readily made quantitative. These latter issues await further research.

APPENDIX: POLITICAL ACTION COMMITTEE ASSIGNMENTS

1. Environmental Groups

Sierra Club
Blue Heron Fund
Back Pac
League of Conservation Voters
  of California, of San Diego,
  and of Oregon
Livable World
Desert Caucus
Enviropac
Friends of the Earth

Friends of the River
New Jersey Environmental Voters
New Mexico Conservation Voters
Virginia Conservation Voters
Connecticut Environmental Voters
Columbia River Trust
Campaign for Clean Air

2. Labor Groups

Northwest Forest Workers Association
United Mine Workers
United Paper Workers
Woodworkers PAC

3. Timber Industry

All firms listed under the headings (based on SIC codes) lumber and wood products, and paper and allied products in *The PAC Directory* (1982).

4. Corporations

All contributions from committees coded as associated with corporations in the Federal Election Commission data.

REFERENCES

Barone, Michael; Grant Ujifusa; and Douglas Matthews. 1984. *Almanac of American Politics: 1984*. New York: E. P. Dutton.

Booth, Douglas E. 1991. "Timber Dependency and Wilderness Selection: The U.S. Forest Service, Congress and the RARE II Decisions." *Natural Resources Journal* 31 (4): 715–39.

Chappell, Henry W., Jr. 1982. "Campaign Contributions and Congressional Voting: A Simultaneous Probit-Tobit Model." *Review of Economic and Statistics,* February, 77–83.

Coates, Dennis, and Michael Munger. 1995. "Legislative Voting and the Economic Theory of Politics." *Southern Economic Journal* 61 (3): 861–72.

Congressional Quarterly Almanac: 1983. Washington, D.C.: CQ Press.

Durden, Garey C.; Jason F. Shogren; and Jonathan I. Silberman. 1991. "The Effects of Interest Group Pressure on Coal Strip-Mining Legislation." *Social Science Quarterly* 72 (2): 239–50.

Enelow, James M., and Melvin J. Hinich. 1984. *The Spatial Theory of Voting: An Introduction*. New York: Cambridge University Press.

Guilkey, David K.; Thomas A. Mroz; and Larry Taylor. 1992. "Estimation and Testing in Simultaneous Equations Models with Discrete Outcomes Using Cross Section Data." Mimeographed.

Hinich, Melvin J., and Walker Pollard. 1981. "A New Approach to the Spatial Theory of Electoral Competition." *American Journal of Political Science* 25 (2): 323–41.

Jackson, John E., and John W. Kingdon. 1992. "Ideology, Interest Group Scores, and Legislative Votes." *American Journal of Political Science* 36 (3): 805–23.

Kalt, Joseph, and Mark Zupan. 1984. "Capture and Ideology in the Economic Theory of Politics." *American Economic Review* 74:279–300.

———. 1990. "The Apparent Ideological Behavior of Legislators: Testing for Principal-Agent Slack in Political Institutions." *Journal of Law and Economics* 33:103–31.

Kau, James, and Paul Rubin. 1979. "Self-Interest, Ideology, and Logrolling in Congressional Voting." *Journal of Law and Economics* 22:365–84.

———. 1981. *Congressmen, Constituents, and Contributors*. Boston: Martinus Nijhoff Publishing.

Morton, Rebecca, and Charles Cameron. 1992. "Elections and the Theory of Campaign Contributions." *Economics and Politics* 4 (1): 79–108.

Pashigian, B. Peter. 1985. "Environmental Regulation: Whose Self-Interests Are Being Protected?" *Economic Inquiry* 23:551–84.

Rivers, D., and Q. Vuong. 1988. "Limited Information Estimators and Exogeneity Tests for Simultaneous Probit Models." *Journal of Econometrics* 39:347–66.

Stigler, George. 1971. "The Theory of Economic Regulation." *Bell Journal of Economics and Management Sciences* 2 (1): 3–21.

Stratmann, Thomas. 1991. "What Do Campaign Contributions Buy? Deciphering Causal Effects of Money and Votes." *Southern Economic Journal* 57:606–20.

———. 1992. "The Effects of Logrolling on Congressional Voting." *American Economic Review* 82 (5): 1162–76.

Weinberger, Marvin, and David U. Greevy. 1982. *The Pac Directory*. Cambridge, Mass.: Ballinger Publishing Co.

Wilderness Society. 1983. *National Forest Planning: A Conservationist's Guide*. 2d ed. Washington, D.C.: The Wilderness Society.

———. 1985. *Protecting Roadless Lands in the National Forest Planning Process: A Citizen Handbook*. Washington, D.C.: The Wilderness Society.

Wilkinson, Charles F., and H. Michael Anderson. 1987. *Land and Resource Planning in the National Forests*. Washington, D.C.: Island Press.

Part 3
The Environmental Bureaucracy

CHAPTER 5

Bureaucratic Discretion in Environmental Regulations: The Air Toxics and Asbestos Ban Cases

George Van Houtven

Introduction

Congressional legislation specifies the direction for regulatory policy, but often it is bureaucratic decisions that ultimately define policy. In effect, regulatory agencies must interpret legislation in order to implement it, and this gives rise to bureaucratic discretion. This chapter is an analysis of bureaucratic discretion in the arena of environmental policy. It examines two sets of regulations that were promulgated by the Environmental Protection Agency (EPA) under separate statutes. These two statutes differ critically in how they intend for cost-benefit considerations to be used in setting regulations. The evidence from EPA's decisions depicts an agency that used substantial discretion in its rule making; however, the evidence also suggests that it did so in opposite directions under the two programs. Moreover, in both cases EPA's decisions were successfully challenged in court by groups on opposing sides of the environmental debate. The rules were remanded to EPA with the instruction that it reconsider them with closer adherence to its legislative mandate. This analysis provides statistical evidence to support the courts' conclusions and, for one of the cases, it examines how the relevant court ruling constrained EPA's discretion in its subsequent regulatory decisions.

The two EPA programs studied here are (1) the 1989 ban on asbestos-containing products and (2) the National Emission Standards for Hazardous Air Pollutants (NESHAPs, also known as the Air Toxics program). The first was promulgated under the Toxic Substances Control Act (TSCA) and the second under the Clean Air Act (CAA). In both cases, EPA has promulgated rules that limit human exposure to airborne carcinogens; however, the toxic substances legislation explicitly requires costs and benefits to be balanced in setting regulations, whereas the statute guiding the air regulations is much more restrictive with regard to costs.

Underlying the analysis of bureaucratic discretion in these two cases is the question: Do EPA's regulatory decisions under the separate programs reveal a balancing of costs and benefits? Furthermore, when they are balanced, at what rate are costs and benefits traded off? The rate at which the agency is willing to trade off increased monetary costs for reductions in cancer risk can be interpreted as the implicit dollar value it attaches to reductions in cancer risk.[1] I estimate these values for each program in order to assess (1) how they differ across programs and (2) how they compare with empirical estimates of the value of a statistical life. For the Air Toxics case, I also examine whether this implicit value changes after a crucial court decision reversed EPA's prior rules. In addition, I examine how the distribution of risk within the exposed population influences the regulatory decisions.

Conceptual Framework

The theoretical foundation for this study is best described as the "revealed preference" approach to regulatory decision making, which was originally applied by McFadden (1975) in his analysis of freeway route selection. It has also been applied in the arena of environmental policy to analyze EPA industrial effluent standards (Magat, Krupnick, and Harrington 1986) and pesticide regulations (Cropper et al. 1992). A similar approach was used by McConnell and Schwarz (1992) to analyze the selection of wastewater treatment technology by local regulators. The basic hypothesis is that there exist discernible underlying decision rules that govern bureaucratic decision making and, although these may be more implicit than explicit, they can be measured statistically. To the extent that the factors in the decision-making process can be quantified, one can calculate policy "weights" to reflect the relative importance of these various factors and compare them across regulatory programs.

The fundamental criteria for agency action (or inaction) are found in the language of the enabling legislation; however, even the most precise legislative guidelines leave some scope for interpretation and agency discretion. Two opposing theories of bureaucratic discretion have evolved in the literature. The agency-dominance view asserts that bureaucracies operate with significant autonomy (Wilson 1980); in effect, regulators are able to insulate themselves from congressional oversight. The congressional-dominance approach challenges this position (Weingast and Moran 1985) by arguing that Congress does have the means to discipline errant bureaucracies. The threat of sanctions and budget cuts creates the necessary incentives for agencies to adhere to congressional intent. Applying revealed preference analysis to the two EPA

---

1. It is important to note that the health benefits for these two programs are never explicitly calculated in monetary terms by EPA; therefore, the term *cost-benefit analysis* is used loosely here and may be more accurately described as cost-effectiveness or cost-risk analysis.

programs offers interesting insight into the degree of bureaucratic discretion in setting environmental regulations. The results of this study depict an agency with significant discretionary scope. Interestingly, the direction in which the agency applies its discretionary authority appears to be quite different from one regulatory program to another. This suggests that the agency has not been guided by a monolithic "objective function" in setting its rules.

While the results do provide evidence of agency discretion, the two programs have also been successfully challenged in court. This illustrates another important avenue for disciplining government agencies. In both circumstances, the courts have found that EPA improperly considered costs in its decisions and, in so doing, exceeded the bounds of its legislative mandate. The Air Toxics case is of particular interest because regulations were promulgated both before and after the court decision. This allows us to test how effectively the court decision restricted EPA's discretion with regard to balancing costs and benefits.

Although EPA may often have substantial leeway for balancing conflicting interests in its regulatory decisions, it is not the agency's practice to state explicitly the rate at which it values a human life or a cancer case avoided. Nevertheless, to the extent that EPA balances risk reductions against the costs of regulations, it is implicitly making this type of valuation. In addition, EPA may often weigh the importance of risks which are high but concentrated in a relatively small population versus those that are lower but more prevalent in society. In other words, the *distribution* of risk[2] in the population may play a crucial role as well in the regulatory process. Neither statute is explicit regarding the role of risk distribution in setting regulations; therefore, this presents another facet in which EPA has significant regulatory discretion. By applying a revealed preference approach to the two EPA programs, it is possible to estimate the agency's implicit valuation of a cancer case avoided and to examine how this implicit value depends (1) on the relevant statute and court decisions and (2) on how the risk is distributed.

The Two Approaches to Regulating Toxic Substances

The Asbestos Ban and the Air Toxics regulations share a common objective: the regulation of toxic substances (in particular carcinogens). However, they differ in two fundamental respects. First, as mentioned before, they are enabled by statutes with different guidelines for cost-benefit considerations.

---

2. In order to account for this, a distinction must be made between population risk, as measured by the number of excess cancer cases, and individual risk, as measured by the excess probability of contracting cancer, due to exposure to a carcinogen. Whereas population risk can be calculated from average individual risk, the distribution of risk is often characterized by the maximum level of individual risk (MIR). Reduction in population risk can thus be characterized as an efficiency measure and reductions in MIR as an equity measure of risk reduction.

Whereas the Toxic Substances Control Act expressly instructs EPA to conduct cost-benefit analyses and to justify its rules accordingly, the Clean Air Act is more restrictive with regard to the role of costs in the rule-making process. Second, their regulatory approaches are quite different. Whereas the Asbestos Ban targets *products* that contain asbestos, the air regulations regulate *production processes* to limit emissions of several toxic substances. For example, whereas air standards were promulgated in the early 1970s to reduce emissions of asbestos fibers in the process of mining, milling, construction, demolition, etc., the Asbestos Ban required the outright prohibition of specific products containing asbestos. The product ban approach, which has been used in other environmental programs (pesticides, PCBs), represents, in many ways, a simpler type of decision than the emission standards approach, which often involves selecting a standard from a broad range of technological alternatives.

We can, nevertheless, analyze the two types of regulatory decisions analogously by assuming that the decision maker is a utility maximizing agent who selects those regulatory options that deliver the highest level of utility and by assuming that there is an underlying utility function that guides these decisions. Although utility cannot be observed or measured, we can attempt to elicit those factors that contribute to utility (the arguments of the function) as well as to measure the relative importance of these factors (the parameters of the function) by applying a simple probit model to the dichotomous choices that make up the Asbestos Ban and a multinomial logit model to the more complicated toxic air pollutant decisions.

Measuring Regulatory Costs and Benefits

In gathering data on costs and benefits, I wish to capture the information available to the agency at the time of each decision. This analysis, therefore, relies primarily on data available in EPA's decision documents and in the Federal Register. In many cases EPA revises its estimates between the time it originally proposes a rule to the time it publishes its decision, and often this is in response to comments or additional information from parties outside the agency. Certain theories of agency behavior (Porter and Sagansky 1976) suggest that external information flows are effectively used by interest groups to influence decisions and, as a result, the information may be distorted. In this analysis, I assume that the information EPA presents in support of its rules represents the decision makers' best approximation of reality. Even if their information is inaccurate or does not properly measure social costs, what is important is that the variables adequately reflect what EPA officials perceived their values to be.

Costs are, of course, measured in dollars; however, health effects do not lend themselves to such a convenient metric. As a result, the regulations studied here have focused on mortality risk and, more narrowly, on the risks of contracting cancer. This is not to say that other health endpoints are ignored, but they are clearly given less emphasis in the regulatory process.

An individual's risk from exposure to an environmental carcinogen is expressed as the incremental probability over one's lifetime of contracting cancer. Because of differences in susceptibility and exposure, risks may vary considerably over the exposed population; nevertheless, based on estimates of these risks and the size of the exposed population, it is possible to estimate the expected number of cancer cases attributable to a specific carcinogen. The benefits of regulation are, therefore, most often expressed as the reduction in cancer incidence (population risk); however, even in cases where cancer incidence is low, certain individuals may be faced with relatively high risk and, for this reason, benefits can also be expressed in terms of reductions in risk to the maximally exposed individual (MEI). This distinction is particularly evident in the Air Toxics rules, but it may also play a role in the regulation of asbestos products.

The Asbestos Ban

In contrast to the Air Toxics decisions, the Asbestos Ban concentrated on only one hazardous substance and was promulgated as a single rule. Because the rule was anticipated to impose costs in excess of $100 million on the economy, a Regulatory Impact Analysis (RIA) was conducted, which evaluated the potential economic and health effects of banning almost 40 different asbestos-containing products. The list consists mostly of automotive, construction, and paper products, all of which have the potential to release asbestos fibers into the air during various stages of their lifecycles. Although other regulatory strategies were considered, such as labeling and "phase-outs" of the products, the focus of the RIA was on the effects of banning the products. Ultimately, EPA decided to ban 27 of the 39 products in three separate stages. Seven of the products were banned in the first stage (1990), eight in the second (1993), and twelve in the final stage (1996), leaving twelve products unregulated.

The primary adverse health effect of exposure to asbestos fibers is lung disease, in particular lung cancer. Where data were available, the RIA modeled annual occupational and nonoccupational exposure in each of the asbestos product markets. Using a thirteen-year time horizon[3] for exposures, EPA calculated the resulting cancer incidence for each product category. Based on these estimates, the reduction in cancer incidence could be esti-

---

3. EPA states in the final rule, "The 13-year time period serves as a reasonable endpoint for the analysis at a point well after all the actions taken in the rule have become effective."(54-FR-29483) They use the same time horizon for estimating the economic effects of the rule.

mated for banning each product at each stage. Data on individual levels of exposure indicate that, whereas the *ratio* of occupational cancer cases to total cancer incidence varies considerably from one product to another, there is a clear pattern of high exposure to relatively few people under occupational scenarios and low exposure to much higher populations in nonoccupational circumstances. Although EPA estimated that "thousands of asbestos workers and members of the general population incur individual risks near 1 in 1,000 . . . and that millions of people incur risks near 1 in 1,000,000"[4] from exposure to these products, they did not report maximum individual risk levels for each of the products. Nevertheless, the RIA did subdivide occupational and nonoccupational exposures into categories such as manufacturing, installation, repair, and disposal, and it estimated the exposure levels (in millions of fibers inhaled per year) for each of these subgroups. In the text of the final rule, EPA often emphasizes high individual risks as a reason for banning a product. These data allow us to empirically test whether risks to the most highly exposed subgroup can independently help to explain the regulatory decisions and whether occcupational exposures are weighted differently from nonoccupational exposures.

In order to estimate the economic impacts (over the same thirteen-year period) of a ban, the RIA modeled the markets for each of the products, so as to measure the consumer and producer surplus losses (the sum of which gives us total costs or "gross total loss") that would result from banning the products at the three different points in time.[5] Sensitivity analyses were performed using different assumptions about the future course of demand for each asbestos product, as well as assumptions about future prices of asbestos fiber and asbestos product substitutes. In the final rule, however, EPA identified its most preferred, or "central case," assumptions to analyze the cost effects of the ban (using a "low decline" baseline for the demand for each asbestos product and a 1 percent annual decrease in substitute product prices).

The agency estimated the total present value of the costs of the rule (using a 3 percent discount rate) to be approximately $460 million, and about 200 cancer cases were expected to be avoided. This translates to $2.3 million

    4. U.S. Environmental Protection Agency. July 12, 1989. "Asbestos: Manufacture, Importation, Processing and Distribution in Commerce Prohibition. Final Rule," *Federal Register* 54:29,480.

    5. It should be noted that the gross total loss associated with banning any one product is not, strictly speaking, independent of actions taken against other products. If one or more products are banned, this reduces demand for and, hence, the price of asbestos fiber. The remaining, non-banned markets benefit as a result; therefore, the economic loss to society must be adjusted downwards to account for this secondary effect. This feedback argument is not ignored in the RIA; however, it will not be accounted for in this analysis because the secondary effects are generally quite small in comparison to the primary effects of banning and including them would greatly complicate the model.

per cancer case avoided; however, it obscures the fact that on a product-by-product basis, the cost per cancer case avoided ranged from less than $500,000 to almost $500 million. It also provides no information on the costs and benefits forgone for those products that were not banned. Applying the model described below gives one a better sense of where EPA was implicitly "drawing the line" on costs per cancer case avoided, and the results indicate that it is at a rate considerably higher than the $2.3 million referred to earlier.

In order to specifically analyze the banning decision, I have abstracted from the problem of the timing of the ban. This was achieved by using the gross total loss and incidence reduction estimates from the middle (second) stage ban. Although actual costs and benefits vary depending on the stage at which the ban was imposed, this allows one to compare costs and benefits across products in a consistent manner.[6]

Table 1 presents these costs and benefits for 31 of the 39 products.[7] Costs per cancer case avoided are also calculated for each of the 31 products. Figure 1 plots the costs of regulating each product against its corresponding incidence reduction for the 21 products that were banned, as well as for the 10 that were not. Not surprisingly, the lower the cost and the higher the number of cancer cases avoided (moving southeast on the graph) the more likely it is that the product is banned; however, the overlap between the banned and not-banned products clearly illustrates that the decision is not based deterministically on the cost per cancer case avoided.

By applying a random utility framework to the banning decision, we can express the decision maker's utility of regulation (the change in utility with respect to the no action alternative) as

$$U_i = (\beta_1 \times COST_i) + (\beta_2 \times CANCER\ INCIDENCE_i)$$

$$+ \beta_3 X_i + \mu_i,$$

where $COST_i$ is the dollar cost of regulating product i and $CANCER\ INCIDENCE_i$ is the reduction in cancer incidence that would result from regulation. X_i is a vector of other factors, for example, maximum individual baseline risk, which may influence the agency's decision with regard to product i. X_i may also include a constant term. The coefficient for a constant would indicate an "institutional bias," in the sense that a positive (negative) value would

---

6. Because costs and incidence reductions fall more or less proportionately (cost-per-cancer-case-avoided remains relatively stable) as one moves from an early to a later ban, the results change very little if one uses the estimates from a different stage.

7. Data are not presented for nine of the 39 products, either because the product was no longer being produced by the time the RIA was prepared or because EPA did not have sufficient information with which to estimate costs and benefits.

TABLE 1. Costs and Benefits of Banning Asbestos-Containing Products

Product Description	Gross Total Loss (millions 89$)	Cancer Cases Avoided	Cost per Cancer Case Avoided (millions 89$)
Products banned in first stage (1990)			
Roofing felt			
A/C sheet, flat	4.04	0.9717	4.16
A/C sheet, corrugated	1.72	0.6752	2.55
Pipeline wrap	0.15	0.0923	1.63
Flooring felt	0.55	1.1196	0.49
V/A floor tile			
Asbestos protective clothing			
Products banned in second stage (1993)			
Automatic trans. components	0.20	0.0004	500.00
Sheet gaskets	85.69	1.99728	42.90
Clutch facings	10.93	0.5444	20.08
Beater-add gaskets	97.94	5.93436	16.50
Friction materials	2.06	0.4719	4.37
Disc brake pads LMV (OEM)	3.49	0.9063	3.85
Disc brake pads HV	0.32	0.1948	1.64
Drum brake linings (OEM)	7.18	7.6476	0.94
Products banned in third stage (1996)			
A/C shingles			
A/C pipe	31.66	0.4111	77.01
Roof coatings	178.53	3.9999	44.63
Millboard	75.63	1.9134	39.53
Non-roofing coatings	5.16	0.7399	6.97
Specialty paper	2.27	0.3833	5.92
Disc brake pads LMV (aftermarket)	0.02	0.033	0.61
Brake blocks	5.69	23.2356	0.24
Drum brake linings (A/M)	2.82	12.9784	0.22
Corrugated paper	13.87	136.3872	0.10
Rollboard			
Commercial paper			
Products not banned			
Asbestos diaphragms			
Missile liner	2,314.75	0.214	10,816.59
Acetylene cylinders	1,001.67	0.3161	3,168.84
Sealant tape	0.08	0.00003	2,666.67
Asbestos thread, yarn, etc.	41.19	0.1115	369.42
Sheet gaskets/PTFE	159.15	0.6222	255.79
High grade electrical paper	31.69	0.22192	142.81
Asbestos-reinforced plastics	58.79	0.5107	115.12
Beater-add gaskets/2	40.58	0.657	61.77
Asbestos packing	50.45	1.04724	48.18
Battery separators	0.49	0.0114	42.98
Arc chutes			

Note: All estimates are for banning in the second stage.

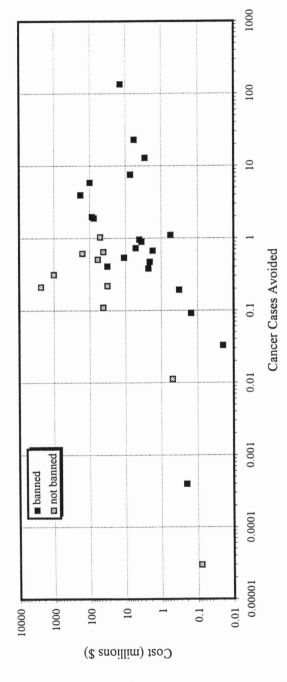

Cancer Cases Avoided

Fig. 1. Cost-effectiveness of asbestos ban

imply a preexisting preference (aversion) by the EPA for regulating. It can also be thought of as a measure of the inherent utility (disutility) of banning, which is independent of the characteristics of the products. The term μ_i is the random component of utility which reflects unobservable characteristics of the *ith* decision and is assumed to be normally distributed with a mean of zero and variance σ. We can, therefore, express the probability of regulation as

$$P(U_i > 0) = P[\mu_i > -(\beta_1 \times COST_i + \beta_2$$

$$\times CANCER\ INCIDENCE_i) + \beta_3 \times X_i]$$

and apply a probit model to estimate the β's.

Dividing both sides of the equation by β_1 effectively monetizes utility, as well as the other variables,

$$U_i/\beta_1 = COST_i + (\beta_2/\beta_1 \times CANCER\ INCIDENCE_i) + (\beta_3/\beta_1)$$

$$\times X_i + \mu_i/\beta_1.$$

The new coefficient on *CANCER INCIDENCE*, β_2/β_1, can therefore be interpreted as the monetary value of a cancer case avoided.

Table 2 shows the results of applying a probit model to the 31 observations. Each equation includes the gross total loss and the cancer incidence reduction as well as a constant term. Because the coefficient on the constant term is insignificant for every specification of the model, there does not appear to be any evidence of an institutional bias in these decisions. On the other hand, the coefficients for costs and benefits are significant, particularly in specifications 3 and 4 where their ratio implies that EPA has valued a cancer case avoided at approximately $49 million (1989 dollars) in this rule. The 95 percent confidence intervals generally place this estimate between $35 and $65 million.

The first specification includes average annual exposure (millions of fibers per year) to the most highly exposed subgroup of individuals. Although the sign of the coefficient supports the hypothesis that high individual risks increase the likelihood of banning (independently of cancer incidence), it is not significant. The third equation uses a similar approach to test the effect of occupational exposure. If the preference for regulation depended directly on the percentage of cancer cases occurring through occupational exposure, this variable would be positive and significant; however, it is not significant either. (This specification was estimated without a constant term because its inclusion creates collinearity.) Equation 2 tests a similar hypothesis by dividing cancer incidence into occupational and nonoccupational categories to see if

TABLE 2. Decisionmaking for Asbestos Ban—Probit Model

Variable	(1)	(2)	(3)	(4)
		Specification		
Constant	−0.67	0.07		0.31
	(−0.7)	(0.1)		(0.6)
Gross total loss (millions	−0.11	−0.17	−0.1	−0.099
1987$)	(−1.7)	(−1.4)	(−2.0)	(−2.0)
Cancer incidence	5.48		4.96	4.85
reduction (no. of cases)	(1.8)		(2.1)	(2.1)
Occupational cancer		11.76		
incidence reduction		(1.4)		
Nonoccupational cancer		5.69		
incidence reduction		(1.4)		
Percentage of cancer				
cases occurring through			0.52	
occupational exposure			(1.0)	
Maximally exposed				
individual's average				
annual exposure	0.004			
(millions of fibers/yr.)	(1.1)			
Log-likelihood	−5.48	−4.91	−6.06	−6.42
Percentage correctly				
predicted	81%	87%	87%	87%
Implicit valuation of a				
cancer case avoided				
(millions 1987$)	47.77	34.39–71.01	48.87	48.61
95% confidcence interval	[36.42, 59.13]		[38.52, 59.22]	[36.66, 60.55]

Note: T-values are in parentheses.

they have a different weight in the regulatory decision. A high degree of correlation between the two incidence measures (0.85) precludes any meaningful tests of significance; however, the coefficient on occupational incidence is almost twice the nonoccupational coefficient which, taken at face value, implies an EPA valuation of an occupational cancer case avoided at about $71 million and a nonoccupational cancer case at $34 million.

The evidence suggests, therefore, that EPA did indeed balance costs and benefits in its banning decisions and that it did so without any predisposition towards banning or not banning. However, considering the language of its statute, it is striking that the implicit value of a cancer case avoided is found to be so high for the Asbestos Ban. After a group of asbestos manufacturers challenged the asbestos rule, the U.S. Court of Appeals for the Fifth Circuit reached a somewhat similar conclusion in October 1991 when it struck down the ban in its entirety. Its assessment of EPA's decisions was even more extreme: "EPA, in its zeal to ban any and all asbestos products, basically ignored the cost side of the TSCA equation."[8] In addition, the court found EPA's use of high individual exposure levels as an additional rationale for banning certain products to be "redundant."[9] Presumably the consideration of risk distribution was also beyond its discretionary scope. Unlike for the Air Toxics, EPA has not promulgated new asbestos regulations since the court decision. Unfortunately this precludes any empirical analysis of how the court decision has affected regulatory decisions under the Toxic Substances Control Act. The next section analyzes Air Toxic regulations and, although the framework is the same, it has the added dimension of being able to assess decisions both before and after a pivotal court decision.

National Emissions Standards for Hazardous Air Pollutants

Compared to the Toxic Substances Control Act, the Clean Air Act is much less explicit about the role of costs in setting air quality standards and regulating emissions of air pollutants. The ambiguity in the legislation's language has only helped to fuel the controversy over EPA's decisions. Much of the criticism has been directed at the National Emissions Standards for Hazardous Air Pollutants, with accusations on one side that the regulations have imposed excessive costs on the economy and concerns on the other that the EPA has exceeded its mandate by factoring costs into the process of setting standards.

Section 112 of the CAA directs the EPA to supplement its National Ambient Air Quality Standards (NAAQS), which control excessive amounts of "criteria" pollutants, with rules to regulate less ubiquitous but more toxic airborne substances such as benzene and mercury. These are the NESHAPs. The procedure defined by section 112 is for EPA to first establish a list of air pollutants that are likely to result in "an increase in mortality or an increase in serious irreversible or incapacitating reversible illness"[10] and, subsequently, to propose emissions standards to protect the public's health with "an ample margin of safety."[11] The ambiguity surrounding the interpretation of this "ample margin of safety" and the appropriate role of costs led EPA to make the following statement in a 1985 response to comments regarding a proposed rule setting standards for emissions of inorganic arsenic:

---

8. *Corrosion Proof Fittings v. EPA*. 1991. 947 F.2d 1218 (5th Cir.), 39.
9. Ibid., 31.
10. 42 U.S.C. 7412(a) (1) (1982).
11. 42 U.S.C. 7412(b) (1) (A)-(B) (1982).

At face value, section 112 could be construed to require regulation even when the costs clearly exceed any measurable benefit. A total disregard for economics would result in a zero risk philosophy. However, this philosophy has been dismissed by EPA. . . . EPA has sought to construct an approach to the implementation of section 112 which . . . considers current control levels and associated health risks as well as options for further control, the health risk reductions obtainable and the associated costs and economic impacts.[12]

In 1987, the Natural Resource Defense Council challenged this position and brought suit against EPA. The U.S. Court of Appeals for the District of Columbia, in what has come to be known as the *Vinyl Chloride* decision,[13] ruled that EPA had improperly considered costs in setting its previous NESHAPs. EPA was directed to consider costs and technological feasibility only once an "acceptable risk" level had been achieved. Against this backdrop of expressed intentions and legal constraints it is interesting to look at the actual record of EPA's decisions to discern its "revealed preferences" or tendencies.

Perhaps because of the considerable controversy and cost involved in the Air Toxics program, rules have been promulgated for only seven substances since the program's inception in the early 1970s—asbestos, benzene, berrylium, inorganic arsenic, mercury, radionuclides, and vinyl chloride—while one has been at the proposal stage for several years—coke oven emissions. In each case, EPA has identified separate source categories of emissions and set standards on a source-by-source basis. Because of the variety of source categories for any one chemical, this has translated into a broad range of regulatory options.

Once chemicals are formally listed under section 112 as hazardous air pollutants, emissions and human exposure estimates are calculated for the specific source categories. For each of these categories EPA considers varying degrees of emissions controls which are achievable with different types or combinations of technologies. Because of the nature of the pollution control technologies, this generally translates into a set of discrete regulatory alternatives whose costs may differ considerably. Therefore, the decision-making procedure for each of these sources is more complex than the dichotomous choice (ban versus do not ban) inherent in the Asbestos Ban. In setting

---

12. U.S. Environmental Protection Agency. 1985. *Inorganic Arsenic NESHAPs: Response to Public Comments on Health, Risk Assessment, and Risk Management* (April).

13. *Natural Resources Defense Council, Inc. v. EPA*, 824 F.2d at 1146 (1987).

emission standards, the agency must select the optimal degree of control from a set of discrete alternatives (which includes the option not to regulate).

The first set of standards was promulgated in 1971 for sources of asbestos, beryllium, and mercury. At this early stage there were no quantified data on risk and only crude measures of cost; therefore, these decisions cannot be included in the statistical analysis. Beginning in 1975, with the rule for vinyl chloride emissions from ethylene dichloride/vinyl chloride and polyvinyl chloride plants, improvements in risk analysis allowed for quantitative estimates of cancer incidence and maximum individual risk (MIR). Table 3 describes the range of baseline risks, where available, for sources of benzene, inorganic arsenic, radionuclides, and vinyl chloride, all of which are considered to be carcinogens by EPA. Because of the 1987 court ruling specifying new procedural guidelines and/or due to new information, some of the sources were reevaluated. For this reason, some of the source categories appear twice in the data set and are treated as distinct observations of regulatory decisions. Figure 2 plots the baseline MIR and cancer incidence attributable to 40 regulated and unregulated sources. Clearly, calculations of risk have played a role in the decision to regulate, as the regulated sources are generally in the upper righthand corner of the graph; however, baseline risks do not entirely explain the decision. The analysis that follows focuses on the 34 categories where costs were calculated, in order to account for their influence. Table 4 presents risk and cost data[14] for those source categories where standards were established. As in the Asbestos Ban, the range in costs per cancer case avoided is quite large, particularly after 1987.

The model used to evaluate these decisions is fundamentally the same as the previous model, except that it incorporates information on the difference in regulatory costs and risk reductions *between alternatives for the same* source as well as *across* sources. In order to do this I use a multinomial/ conditional logit model. As before, I assume that the chosen alternative generates the highest expected utility. If utility is a function of costs and benefits, then the difference in utility between any two regulatory alternatives (alternative j and alternative k) can be expressed as:

$$U_{jk} = (\beta_1 \times COST_{jk}) + (\beta_2 \times RISK_{jk}) + \epsilon_{jk},$$

where U_{jk} is the difference in utility, $COST_{jk}$ is the difference in cost, and $RISK_{jk}$ is the difference in the level of risk that would result from each

---

14. In the NESHAPs case, only compliance costs (capital and operation and maintenance) were estimated, which may not be the same as the true economic costs of the standards considered.

TABLE 3. Baseline Risks for Sources of Air Toxics

Substance	Source	Decision Year	Maximum Individual Risk	Annual Cancer Incidence
Regulated sources				
Benzene	Benzene transfer operations	90	6.00E-03	1
Benzene	Benzene waste operations	90	2.00E-03	0.6
Benzene	Benzene storage vessels	89	1.30E-04	0.071
Benzene	Coke by-product recovery plants	89	7.00E-03	2
Radionuclides	Elemental phosphorous plants	89	5.70E-04	0.072
Radionuclides	Radon releases from DOE facilities	89	1.40E-03	0.072
Radionuclides	Underground uranium mines	89	4.40E-03	0.79
Radionuclides	Operating uranium mill tailings (new)	89	1.60E-04	0.014
Arsenic	Primary copper smelters	86	1.30E-03	0.38
Arsenic	Glass manufacturing plants	86	9.00E-04	0.4
Benzene	Benzene fugitive emissions (existing)	82	1.46E-03	0.45
Benzene	Benzene fugitive emissions (new)	82	1.46E-03	0.12
Vinyl Chloride	EDC/VC and PVC plants	75	4.86E-03	11
Unregulated sources				
Benzene	Bulk gasoline terminals	90	5.00E-05	0.12
Benzene	Bulk gasoline plants	90	1.00E-05	0.03
Benzene	Service station storage vessels	90	5.00E-06	0.13
Benzene	Rubber tire manufacturing (ISU)	90	4.00E-06	0.0006
Benzene	Pharmaceutical manufacturing	90	1.00E-06	0.001
Benzene	Chemical manufacturing process vents	90	4.00E-06	0.01
Benzene	Ethylbenzene/Styrene process vents	89	2.00E-05	0.003
Benzene	Benzene equipment leaks	89	1.00E-04	0.2
Radionuclides	DOE facilities	89	2.00E-04	0.28
Radionuclides	NRC licensed & non-DOE facilities	89	1.60E-04	0.16
Radionuclides	Uranium fuel cycle facilities	89	1.50E-04	0.1
Radionuclides	Coal-fired utility boilers	89	2.50E-05	0.4
Radionuclides	Coal-fired industrial boilers	89	7.00E-06	0.4
Radionuclides	Phosphogypsum stacks	89	9.10E-05	0.95
Radionuclides	Surface uranium mill tailings	89	4.80E-05	0.026
Radionuclides	Disposal of uranium mill tailings piles	89	3.00E-04	0.07
Radionuclides	High level nuclear waste disposal facilities*	89	7.00E-08	4.30E-06
Radionuclides	Operating uranium mill tailings (existing)*	89	2.90E-05	4.30E-03
Arsenic	Secondary lead plants	86	4.00E-04	0.39
Arsenic	Chemical manufacturing plants*	86	4.00E-06	0.2
Arsenic	Primary zinc smelters*	86	4.00E-06	0.007
Arsenic	Cotton gins*	86	5.00E-04	n/a
Arsenic	Zinc oxide plants*	86	1.00E-04	0.08
Arsenic	Primary lead smelters*	86	2.00E-04	0.07
Benzene	Maleic anhydride plants	84	7.60E-05	0.029
Radionuclides	Elemental phosphorous plants	84	1.00E-03	0.058
Radionuclides	Coal-fired utility boilers	84	1.00E-05	1.4
Radionuclides	Coal-fired industrial boilers	84	1.00E-06	1

*Costs of regulating source not estimated—not included in data for logit model.

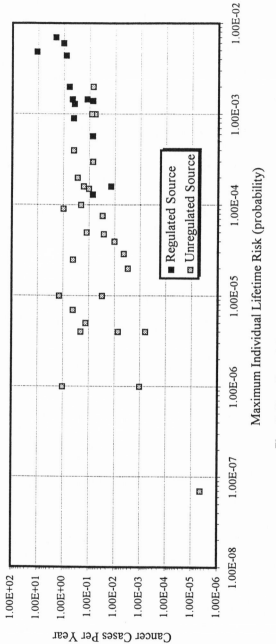

Maximum Individual Lifetime Risk (probability)

Fig. 2. Baseline risks at sources of air toxics

TABLE 4. Risk Reductions and Costs for Regulated Sources of Air Toxics

Substance	Emission Source	Decision Year	Annual Cancer Incidence Reduction	Annual Cost (millions 1989 $)	Modeled Population (millions)	Annual Cost per Cancer Case Avoided (millions 1989 $)
Benzene	Benzene transfer operations	1990	0.9800	32.70		33.37
Benzene	Benzene waste operations	1990	0.5500	98.31		178.75
Radionuclides	Elemental phosphorous plants	1989	0.0480	2.40	1.9	50.00
Radionuclides	Radon releases from DOE facilities	1989	0.0320	1.50	28.4	46.88
Radionuclides	Underground uranium mines	1989	0.5500	0.40	1.2	0.73
Radionuclides	Operating uranium mill tailings	1989	0.0050	0.50	0.8	100.00
Benzene	Benzene storage vessels	1989	0.0310	0.13	70.0	4.13
Benzene	Coke by-product recovery plants	1989	1.9500	19.04	70.0	9.76
Arsenic	Primary copper smelters	1986	0.0900	0.49	1.9	5.40
Arsenic	Glass manufacturing plants	1986	0.3300	4.07	11.6	12.33
Benzene	Benzene fugitive emissions (existing)	1982	0.3100	0.68	25.0	2.21
Benzene	Benzene fugitive emissions (new)	1982	0.0820	0.17	25.0	2.09
Vinyl Chloride	EDC/VC and PVC plants	1975	10.4000	149.1	4.6	14.34

alternative. ϵ_{jk} is a composite stochastic error term (see the Appendix for more detail).

If we denote the chosen alternative as alternative l, this implies that the difference in utility between alternative l and any other alternative is positive. Formally,

$U_{lj} > 0$ for all alternatives $j \neq 1$.

Therefore, for any given source category, the *ex ante* probability that alternative l is chosen can be written as

$$P(\beta_1 \times COST_{lj} + \beta_2 \times RISK_{lj} > -\epsilon_{lj}) \text{ for all alternatives } j \neq 1.$$

By assuming a specific probability distribution for ϵ_{lj}, the conditional logit model takes, for each source category, the differences in costs and risks between the chosen alternative and all other alternatives and uses a maximum likelihood method to estimate the β's.

Table 5 shows the results for various specifications of the model. Equations 1A and 1B focus simply on increases in annual costs and reductions in risk, specifically population risk (annual cancer incidence) and MIR. Unfortunately the results are not statistically significant when both incidence and MIR reductions are included. When only incidence and cost are included (the same specification as the simple asbestos model), both coefficients are significant (at a 0.05 significance level) and imply that EPA has valued a cancer case avoided at about $152 million dollars for this set of rules. This value, however, is somewhat misleading because it does not account for changes due to the 1987 court ruling.

The second set of equations takes into account the fact that 24 of the 34 source control decisions were made after the 1987 *Vinyl Chloride* decision. The basic conclusions of this court ruling were that EPA had improperly weighed costs and technological considerations in its previous decisions and that a determination of "acceptable risk" should be the primary regulatory criterion. In specifications 2A and 2B, costs are interacted with a dummy variable (POSTVC) which is set equal to one when a decision occurred after the 1987 ruling. Again, the results are not very significant when MIR is included in the specification; however, in general, they do indicate that costs are given considerably less weight after 1987.

A Wald test was conducted in each case to test the null hypothesis that costs are given zero weight in the decisions made after 1987. This hypothesis can be rejected at a significance level between 0.03 and 0.06, indicating that costs are still a factor, albeit a less important one. Taking into account this distinction, it appears that the agency valued a cancer case avoided at between

TABLE 5. Decision Making for NESHAPs—Conditional Logit Model

| | Equation | | | | |
Variable	1A	1B	2A	2B	3
Reduction in MIR (×1000)	2.4		1.39		
	(1.3)		(0.9)		
Reduction in cancer incidence	4.2	9.9	18.46	21.64	21.67
	(0.6)	(1.9)	(2.0)	(2.3)	(2.1)
Increase in annual cost (millions 1989 $)*	-0.073	-0.065	-1.26	-1.335	-1.47
	(-2.2)	(-2.4)	(-1.9)	(-1.6)	(-2.0)
Increase in annual cost*			1.14	1.223	
			(1.9)	(1.5)	
Increase in Annual Cost*†					1.37
					(2.0)
Log-likelihood	-17.7	-18.8	-14.1	-14.5	-11.8
Implicit value of a cancer case avoided					
1975–90	57.50	152.64			
[95% confidence interval]		[52.07, 252.94]			
1975–87			14.65	16.21	14.73
[95% confidence interval]			[12.1, 17.2]	[2.22, 30.2]	[10.6, 18.84]
1987–90			153.83	194.06	216.70
[95% confidence interval]			[52.57, 255.09]	[23.93, 264.1]	[80.12, 353.32]

Note: T-values are in parentheses.
*Post 1987 Dummy.
†MIR > .0001 Dummy.

$14 million and $16 million before 1987 and between $154 million and $193 million after 1987. Not surprisingly, the initial estimate of $152 million falls between these two ranges.

A final specification was estimated to more accurately reflect the intent of the *Vinyl Chloride* decision and to account for the concept of "acceptable risk." In 1988, the agency proposed various "policy approaches" in an attempt to define acceptable risk,[15] and, despite the fact that none of the approaches was officially sanctioned, they did propose to use levels of MIR (such as 10^{-4} and 10^{-6}) as benchmarks.

An implication of using a risk benchmark, as intended in *Vinyl Chloride*, is that costs should be given more weight only when the benchmark has been achieved. In other words, EPA should value (in dollar terms) reductions in cancer incidence more when maximum individual risks are high. Using an MIR of 10^{-4} as the benchmark, we can formalize this by writing the difference in expected utility between any two alternatives, i and j (where alternative i is the less stringent alternative), as

$$E(U_{ij}) = (\beta_1 \times COST_{ij})$$

$$+ (\beta_2 \times RISK_{ij}) \text{ if MIR resulting from } i > 10^{-4},$$

and $E(U_{ij}) = (\beta_1 - \delta) \times COST_{ij}$

$$+ (\beta_2 \times RISK_{ij}) \text{ if MIR resulting from } i \le 10^{-4},$$

where $0 > \beta_1 > \beta_1 - \delta$.

In order to incorporate this into the model, we interact the cost variable with the post-1987 dummy as before, and also with a dummy variable that is only equal to one if the MIR that is achieved by the *less stringent alternative* is greater than 10^{-4}.

The coefficients estimated from this specification are all significant at a 0.05 level and imply (1) that a cancer case avoided is valued at approximately $15 million before the 1987 court decision and the same after 1987 if MIR is not greater than 10^{-4} and (2) a cancer case avoided is valued at over $200 million if MIR is above 10^{-4}. A Wald test was again conducted to test the significance of the second result. The null hypothesis is that costs are not weighted in the decision after 1987 if the MIR is greater than 10^{-4} ($\beta_1 = 0$). The null hypothesis can only be rejected at a significance level of 0.11. We

15. U.S. Environmental Protection Agency. July 28, 1988. "NESHAPS, Proposed Rule for Sources of Benzene," *Federal Register* 53:28,497.

therefore cannot reject the possibility that costs are not a factor if MIR exceeds 10^{-4} after 1987.

It appears, therefore, that the *Vinyl Chloride* decision did have a significant impact on EPA discretion with regard to how it weighed cost estimates in setting emission standards. Whereas EPA's decisions reveal relatively strong cost aversion prior to 1987, this tendency is insignificant after 1987 as long as risks are "excessive." Also, there is no evidence that *reductions* in MIR played an independent role in the decisions; however, the *level* of MIR does appear to have been particularly crucial in the post *Vinyl Chloride* decisions in so far as it defines acceptable risk.

Conclusions

The evidence from the Asbestos Ban and the Air Toxics program indicates that, for the most part, EPA's cost and benefit estimates are significant variables in explaining its regulatory decisions. Larger risk reductions increase the probability of regulation, while larger costs lower the probability. Given the statutory language of the Toxic Substances Control Act, this result is not particularly surprising for the Asbestos Ban. On this level, EPA appears to have acted squarely within the constraints of its legislative mandate. On the other hand, similar evidence for the Air Toxics program depicts a more independent-minded regulatory agency. This result is particularly striking when the implicit valuation estimates for cancer cases avoided are compared for the two programs. The implied value of a cancer case avoided is considerably higher in the asbestos case (where EPA is required to balance costs and benefits) than in many of the Air Toxics decisions ($50 million versus $15 million). It is only through the court's *Vinyl Chloride* decision that costs are deemphasized in the regulatory process for Air Toxics and that EPA's regulatory discretion is reined in.

Equally important is how the implicit valuations compare with empirical measures of the value of a statistical life. Fisher, Chestnut, and Violette (1989) surveyed the empirical literature on individuals' willingness to pay (WTP) for reductions in mortality risk and concluded that the most defensible estimates of the value of a statistical life fall below $10 million. They settled on a range between $1.6 million and $8.5 million (in 1986 dollars). By contrast, this study has estimated EPA's implicit valuations of a cancer case avoided under two programs, and they are all above $10 million. The valuation of a death avoided (statistical life saved) would be still greater because not all cancer cases are fatal.[16] Although the WTP studies are not entirely conclu-

---

16. Cure rates depend on the specific type of cancer, but for lung cancer it is about 8 percent (Mauskopf 1987).

sive, and some may argue that they produce underestimates, the results here indicate that EPA's implicit valuation of a life saved has been, if anything, relatively high. This is particularly true for the Asbestos Ban and is a separate indication of EPA using its discretionary authority—although costs and benefits appear to be balanced as required, costs are given relatively little weight.

More striking, perhaps, is that EPA appears to have used its discretionary authority in quite different directions under the two programs. Congleton (1982) argued that bureaucratic discretion may exist and be used in different ways at different levels of the organizational hierarchy. Here we have an example of how this may be true in separate divisions of the bureaucratic organization as well. The difference in discretionary direction across EPA programs may be partially explained by the fact that the institutional structure of decision making at EPA is quite different for toxic substances than for air pollutants—the process is generally more interactive (between those who do the analyses and those who interpret them and make decisions) in the former case. It is not clear *a priori,* however, how this might influence the rate at which costs and risk reductions are traded off in the regulatory decisions. The difference in hierarchical structure within the separate divisions themselves may nevertheless help to explain the different discretionary outcomes.

In its two decades of existence, EPA has seen its freedom to interpret environmental legislation greatly restricted by the courts (see Marcus in Wilson 1980). The regulations examined in this study are certainly not the only cases in which the EPA has been sued by interested parties and had its decisions overturned. However, these cases are distinct because in both cases EPA's decisions were reversed largely on cost-benefit grounds. While debate continues over the usefulness and validity of cost-benefit analysis in environmental regulation, EPA's role in the debate has been eroded. As long as it must respect statutory guidelines that require separate and conflicting criteria, it is hard to imagine that EPA will itself be able to develop a coherent approach for using cost-benefit analysis.

APPENDIX

For multinomial decisions involving two or more alternatives, we can express the utility of regulatory alternative i as

$$U_i = (\beta_1 \times COST_i) + (\beta_2 \times RISK_i) + \epsilon_i, \qquad \text{A(1)}$$

where $COST_i$ is the dollar cost of alternative i and $RISK_i$ is the *level* of risk (whether cancer incidence or MIR) that would result from selecting alternative i. In order to apply the multinomial/conditional logit model to this framework, the random error term, ϵ_i, is assumed to have the type I extreme-value distribution. The probability that alternative i is preferred to any other alternative j can therefore be expressed as

$$P(U_{ij} > 0) = P[\epsilon_{ij} > -(\beta_1 \times COST_{ij}) + (\beta_2 \times RISK_{ij})]. \qquad \text{A(2)}$$

where $U_{ij} = U_i - U_j$, $COST_{ij} = COST_i - COST_j$, $RISK_{ij} = RISK_i - RISK_j$, and $\epsilon_{ij} = \epsilon_i - \epsilon_j$. The probability that alternative i selected is therefore $P(U_{ij} > 0)$ for all $j \neq i$.

If *RISK* is a measure of cancer incidence then, for the same reasons discussed above, we can interpret β_2 / β_1 as the implicit valuation of a cancer case avoided.

REFERENCES

Asch, Peter, and Seneca, Joseph J. 1989. "Determinants of Health and Safety Regulation." Rutgers University Department of Economics. New Series Working Paper no. 1989-06.

Congleton, Roger. 1982. "A Model of Asymmetric Bureaucratic Inertia and Bias." *Public Choice* 39:421–25.

Cropper, Maureen L.; Evans, William N.; Berardi, Stephen J.; Ducla-Soares, Maria M.; and Portney, Paul R. 1992. "The Determinants of Pesticide Regulation: A Statistical Analysis of EPA Decision Making." *Journal of Political Economy* 100 (January): 175–97.

Fisher, Ann; Chestnut, Lauraine G.; and Violette, Daniel M. 1989. "The Value of Reducing Risks of Death: A Note On New Evidence." *Journal of Policy Analysis and Management* 8 (Winter): 88–100.

Graham, John D., and Vaupel, James W. 1981. "Value of a Life: What Difference Does It Make?" *Risk Analysis* 1:89–95.

Haigh, John A.; Harrison, David, Jr.; and Nichols, Albert L. 1984. "Benefit-Cost Analysis of Environmental Regulation: Case Studies of Hazardous Air Pollutants." *Harvard Environmental Law Review* 8:395–434.

McConnell, Virginia D., and Schwarz, Gregory E. 1992. "The Supply and Demand for Pollution Control: Evidence from Wastewater Treatment." *Journal of Environmental and Economic Management* 23:54–77.

McFadden, Daniel. 1975. "The Revealed Preferences of a Government Bureaucracy: Theory." *Bell Journal of Economics and Management Science* 6 (Autumn): 401–16.

———. 1976. "The Revealed Preferences of a Government Bureaucracy: Empirical Evidence." *Bell Journal of Economics and Management Science* 7 (Spring): 55–72.

Maddala, G. S. 1983. *Limited Dependent and Qualitative Variables in Econometrics*. New York: Cambridge University Press.

Magat, Wesley; Krupnick, Alan J.; and Harrington, Winston. 1986. *Rules in the Making: A Statistical Analysis of Regulatory Agency Behavior*. Baltimore: Johns Hopkins University Press.

Mauskopf, Josephine A. 1987. "Projections of Cancer Risks Attributable to Future Exposure to Asbestos." *Risk Analysis* 7:477–86.

Nichols, Albert L. 1991. "Comparing Risk Standards: The Superiority of a Benefit-Cost Approach." *Regulation* (Fall): 85–94.

Porter, Michael D., and Jeffrey F. Sagansky. 1976. "Information Politics and Economic Analysis: The Regulatory Decision Process in the Air Freight Cases." *Public Policy* 24 (Spring): 263–307.

Travis, Curtis; Richter, Samantha A.; Crouch, Edmund A.; Wilson, Richard; and Klema, Ernest D. 1987. "Cancer Risk Management: A Review of 132 Federal Regulatory Decisions." *Environmental Science and Technology* 21 (May): 415–20.

United States Environmental Protection Agency. 1989. "Regulatory Impact Analysis of Controls on Asbestos and Asbestos Products: Final Report." Washington, D.C.: Office of Toxic Substances.

Weingast, Barry R., and Moran, Mark J. 1985. "Bureaucratic Discretion or Congressional Control? Regulatory Policymaking by the Federal Trade Commission." *Journal of Political Economy* 91:765–800.

Wilson, James Q., ed. 1980. *Bureaucracy: What Government Agencies Do and Why They Do It*. New York: Basic Books, Inc.

CHAPTER 6

The Determinants of Pesticide Regulation: A Statistical Analysis of EPA Decision Making

Maureen L. Cropper, William N. Evans,
Stephen J. Berardi, Maria M. Ducla-Soares,
and Paul R. Portney

When asked how standards should be set in environmental, safety, and health regulation, virtually all economists would urge that at least some account be taken of economic factors. Most would probably support the view that such standards should be set at levels that equate marginal social benefits and costs. This approach does not command overwhelming support when legislation is written, however. In fact, U.S. environmental policy could be termed schizophrenic with respect to the balancing of benefits and costs in standard setting: most major statutes appear to *prohibit* such balancing, with the Clean Air and Clean Water Acts perhaps being the most prominent examples; however, other important environmental laws require that benefits and costs be balanced when decisions are made, this being the case with the Toxic Substances Control Act and also the Federal Insecticide, Fungicide, and Rodenticide Act (FIFRA), the latter being the statute under which most U.S. pesticide regulation is conducted.

What laws require is one thing; what agencies do is another. In environmental regulation, for example, White (1981) has argued that although the Clean Air Act has been construed by courts to prohibit consideration of costs in setting ambient air quality standards (*Lead Industries Assoc., Inc. v. EPA* (1980)), EPA has in fact taken economics into account in setting such standards for common pollutants. Similarly, for a time EPA explicitly balanced health risks against economic costs in regulating certain carcinogenic air pollutants. Others have argued, however, that even when the relevant statutes require the balancing of economic and health considerations, agencies will always take action against cancer risks that exceed certain statistical thresh-

Reprinted from *Journal of Political Economy* 100, no. 1 (1992). © 1992 by The University of Chicago. All rights reserved.

olds, often referred to as "bright lines" (Travis et al. (1987), Travis and Hattemer-Frey (1988), Milvy (1986)), regardless of costs.

Finally, still others maintain that no matter what "objective" factors the statutes direct regulatory agencies to consider, the latter are sure to be influenced in their rulemaking in important and predictable ways by political considerations (Stigler (1971), Peltzman (1976)). This view is of particular interest to us in light of the history of pesticide regulation, the focus of our attention here. Prior to the creation of EPA in 1970, all pesticides were regulated by the Department of Agriculture. One of the reasons for transferring regulatory responsibility to EPA was to lessen the influence of farmers and pesticide manufacturers in the regulatory process and increase the influence of environmental and consumer groups (Bosso (1987)).

Although there is a substantial literature on the determinants of legislative voting on environmental issues (see Crandall (1983), Pashigian (1985), Yandle (1989), and Hird (1990)), there exists but one published analysis of EPA decisionmaking to ascertain, ex post facto, the factors which explain the regulatory actions taken (see Magat, Krupnick, and Harrington (1986)).[1] This paper presents such an analysis for a particular class of environmental regulations, viz., EPA's decisions to allow or prohibit the continued use of certain pesticides on food crops. We are interested in whether the economic benefits that pesticides confer are, in fact, balanced against the risks these substances may pose to human health and the environment. We also examine the extent to which these decisions are affected by the active involvement of special interest groups—on the one hand, the companies that manufacture pesticides and the farmers that use them, and, on the other, the environmental advocacy organizations that often oppose the widespread application of pesticides.

We focus on three specific hypotheses. First, is the probability that EPA will disallow continued use of a pesticide on a particular crop positively related to the risks that pesticide poses to human health and the environment and negatively related to the economic benefits associated with the use of the pesticide? In other words, does EPA follow its congressional mandate under FIFRA? If both factors are taken into account by EPA, what is the implicit "price" of the resulting risk reductions? This question is important because of concern that the cost-per-life-saved as revealed in health and safety regulation differs markedly, both within and across agencies (Morrall (1986)); this *may* signal an inefficient allocation of resources amongst life-saving programs.[2]

---

1. For analyses of decisionmaking at other government agencies, see McFadden (1975, 1976), Weingast and Moran (1983), and Thomas (1988).

2. If individuals attach higher values to the reduction of certain kinds of risks—for example, involuntary versus voluntary—or if certain regulations save more life-years than others, variations in cost-per-life-saved may be perfectly rational.

Second, do special interest groups—both business and environmental—affect the likelihood that certain pesticide uses will be banned? If so, when opposite sides both intervene, do their efforts merely offset one another?

Third, can particular political appointees influence the likelihood of regulatory action? During the period covered by our study, EPA was headed by five different administrators including Anne Burford, a Reagan appointee widely regarded as unsympathetic to environmentalists' concerns. We test the hypothesis that she had a significant impact on pesticide regulation during her tenure.

We have investigated these questions by assembling data on all cancer-causing pesticides that underwent Special Review by EPA between 1975 and 1989. Under FIFRA, the Special Review process is initiated whenever a pesticide is thought to pose a danger to human health (e.g., cancer or adverse reproductive effects) or to wildlife; this review entails a risk-benefit analysis of the pesticide for each and every crop on which it is used. Following this analysis, EPA issues a proposed decision and invites all interested parties to submit comments which are compiled in a public docket. A final decision (or Notice of Final Determination) is issued after the Agency has reconsidered its proposed action in light of these comments and any new information it has developed. We have assembled data on the risks and benefits associated with each pesticide from official data published by EPA, as well as information on which special interest groups entered comments in the public docket.

These data are used to estimate a model that explains the probability that a pesticide was cancelled for use on a particular crop, as a function of the risks and benefits associated with its use and as a function of political variables. Our findings provide both comfort and concern to those interested in improving the efficiency of environmental regulation.

I. An Overview of EPA's Pesticide Registration Process

In its 1972 amendments to FIFRA, Congress required EPA to reregister the approximately 40,000 pesticides previously approved for sale in the U.S. In the 1978 amendments to FIFRA, this task was simplified by requiring reregistration of the 600 active ingredients used in these pesticides, rather than the pesticides themselves.

Reregistration of each active ingredient requires assembling the data necessary to evaluate whether it causes "unreasonable adverse effects on the environment" for each use for which it is registered. By "use" is meant the application of a pesticide to a specific crop (e.g., alachlor on soybeans). If, in the process of collecting this data, it is determined that the active ingredient poses sufficient risks to humans or animals, it is put through the Special Review process. The purpose of the process is to determine whether the risks posed by the active ingredient are outweighed by the benefits of its use.

The results of these risk-benefit analyses are published along with EPA's proposed regulatory decision. The following regulatory outcomes are considered for each use of the active ingredient: (1) cancellation of registration; (2) suspension of registration; (3) continuation of registration, subject to certain restrictions; (4) unrestricted continuation of registration.

Publication of the proposed decision is followed by a comment period, during which members of the public, including growers, public interest groups, and registrants, can respond. If cancellation or restrictions on use are contemplated, the U.S. Department of Agriculture and EPA's Scientific Advisory Panel are asked to review the risk-benefit analyses. Final regulatory decisions, together with the names of all those who commented on the proposed decision, are then issued, and these decisions become law unless a hearing is requested by interested parties.

Between 1975, when the Special Review process was initiated, and 1989, a total of 68 Special Reviews were begun (U.S. Environmental Protection Agency, 1989). Of these, 18 ended at a pre–Special Review stage, 37 had been completed by December of 1989, and 13 were ongoing, as of that date. Our study focuses on a subset of the 37 substances for which reviews were completed, viz., those that both involve pesticides used on food crops and have been found to cause cancer in laboratory animals. We focus on this subset because health risks other than risk of cancer are seldom quantified, which makes a statistical analysis of regulatory decisions difficult.

The set of food-use pesticides causing cancer in laboratory animals that have gone through Special Review is listed in table 1. Note that although there are only 19 such pesticides, there were 245 separate pesticide/crop combinations or uses. What we shall try to explain is the decision to cancel or not cancel each of these uses.[3]

II. Factors Influencing the Cancellation Decision

Risks of Pesticide Use

In deciding whether a pesticide should be cancelled for use on a crop, EPA is required to "prevent any unreasonable risk to man or the environment, taking

---

3. Of the 245 final decisions in our database, 39% represent cancellations, 4% suspensions of registration for failure to provide data, 5% unrestricted continuations, and 52% continuations with restrictions. The types of restrictions typically imposed consist of measures to protect pesticide mixers and applicators, such as requiring that protective clothing be worn. These decisions are to be made by comparing the risks and benefits of the restrictions; however, the documents EPA develops typically do not contain enough data to permit an analysis of each restriction. For this reason we consider only two regulatory outcomes: continuation of registration (with or without restrictions) or cancellation. Suspensions for failure to provide data are grouped with continuations, since registrations are continued as soon as the data are provided.

TABLE 1.   Active Ingredients in the Pesticide Data Base

Active Ingredient	Year of Decision	Number of Food-Use Registrations	Number of Proposed Cancellations	Number of Final Cancellations
DBCP	1978	12	1	12
Amitraz	1979	2	1	1
Chlorobenzilate	1979	3	2	2
Endrin	1979	8	4	4
Pronamide	1979	4	0	0
Dimethoate	1980	25	0	0
Benomyl	1982	26	0	0
Diallate	1982	10	10	0
Oxyfluorfen	1982	3	0	0
Toxaphene	1982	11	7	7
Trifluralin	1982	25	0	0
EDB	1983	18	4	18
Ethalfluralin	1983	3	0	0
Lindane	1983	8	7	0
Silvex	1985	6	6	6
2, 4, 5-T	1985	2	2	2
Dicofol	1986	4	4	0
Alachlor	1987	10	3	0
Captan	1989	65	65	44
Totals		245	116	96

into account the economic, social, and environmental costs and benefits of the use of [the] pesticide." Paramount among these risks is the risk of cancer to persons who mix and apply pesticides and to consumers who ingest pesticide residues on food.[4] Evidence that a chemical is carcinogenic usually comes from animal bioassays, which produce a relationship between pesticide dose and lifetime risk of cancer. This estimate is extrapolated to humans and multiplied by an estimate of human dosage (exposure) to estimate lifetime risk of cancer to a farmworker or consumer.[5]

Lifetime cancer risks are typically much higher for pesticide applicators than for consumers of food products; for example, in our sample the median estimated incremental lifetime cancer risk for pesticide applicators (as a result

4. In its official documents, EPA lists cancer risks to pesticide applicators and to persons who mix and load pesticides (mixer/loaders), but not to farmworkers who harvest crops. Risks to farmworkers are controlled by adjusting the time between pesticide application and harvest (the pre–harvest interval).

5. It is well-known that the methods used to calculate the slope of the dose response function, and those used to estimate human exposure to the pesticide, generally err on the side of conservatism (Nichols and Zeckhauser, 1986). The slope of the dose-response function is the upper bound of a 95-percent confidence interval, rather than the midpoint.

of applying a particular pesticide to a particular crop) is 1 in 100,000 (1.0×10^{-5}), but it is only 2.3 in 100 million (2.3×10^{-8}) for consumers of food products.[6] The number of persons assumed to be exposed to dietary risks—usually the entire U.S. population—is, however, much greater than the number of applicators exposed to pesticides. The latter may range from a few dozen to a few thousand, depending on the particular crop and the number of acres treated, while the number of persons mixing pesticides is typically a few hundred.

This raises a very difficult regulatory issue: Should EPA's decisions be driven by very high risks to certain individuals (the so-called "maximally exposed individuals") or by the overall risk to the entire exposed population (that is, the expected number of deaths)? Although economists have typically emphasized the latter, regulatory officials at EPA and other agencies are often more preoccupied with reducing very high individual risks to acceptable levels.

In addition to cancer risks, pesticides may have adverse reproductive effects—causing fetal deformities, miscarriages, or lowering the sperm counts of applicators. While there is human evidence for the latter effects, information on the mutagenic or teratogenic effects of a chemical usually comes from animal experiments, and the extent of such effects is generally difficult to quantify. Finally, EPA is required to consider the possibly adverse ecological effects of pesticides—is the pesticide toxic to fish, birds, or wildlife, or is it likely to contaminate ecologically fragile environments such as wetlands?

Benefits of Pesticide Use

Against these risks, EPA must weigh the benefits of use, i.e., the costs to consumers and producers of banning the pesticide on the crop in question. Losses accrue if producers must switch to a more costly substitute for the pesticide in question or if the substitute is an imperfect one and yield losses will occur upon cancellation. Decreases in supply may, in turn, lead to price increases to consumers.

Losses to producers from cancellation vary widely for the pesticides and crops studied here. The highest loss expected during the first year following cancellation is $227 million (1986 dollars) for alachlor on corn. Mean first-year losses, however, are considerably lower—only $9.1 million. In 35% of all cases losses are negligible, due to the availability of substitute pesticides. What is likely to be as important as the magnitude of losses is their distribution among growers. A 0.1% reduction in corn revenues will greatly exceed

---

6. To put this in perspective, we note that the average lifetime cancer risk from all causes is one-third.

that associated with a 50% decrease in mango production; however, since there are relative few mango growers, the distribution of losses is far more concentrated in the latter case than in the former.

The Role of Political Factors

This raises directly the question of the importance of interest groups in the regulatory process. Pesticide manufacturers are, of course, involved throughout—they are informed when EPA contemplates a Special Review and are given an opportunity to rebut the presumption that the pesticide causes adverse effects to humans or to the environment. In addition to negotiating with EPA, manufacturers are responsible for providing data on the risks of pesticide usage.

Farmers also bear the costs of cancellation and thus have an interest in dissuading EPA from banning pesticides. One would expect farmers to become involved when the cost of switching to substitute pesticides is high, and when the losses that would result constitute a large percentage of profits. An interesting question is at what stage in the regulatory process farmers become involved. While anecdotal evidence suggests communication between EPA and grower organizations throughout the regulatory process, farmers have no need to exert leverage unless they feel that a pesticide is threatened with cancellation. Thus, one would expect grower organizations or their representatives to comment more often when EPA proposes to cancel rather than to allow continued use of the pesticide(s) in question.

Environmental groups, which attempt to identify and fight for the cancellation of pesticides hazardous to humans and/or wildlife, can be expected to behave differently. They may, moreover, exert an influence earlier in the regulatory process by bringing pesticide risks to EPA's attention before the official comment period. Finally, one would expect the views of the EPA Administrator to affect the outcome of the Special Review Process since it is the job of the Administrator to review the evidence on health and environmental risks, and the economic effects associated with the cancellation decision and to issue a final decision.

III. Statistical Analysis of EPA's Pesticide Decisions

If EPA follows its mandate under FIFRA to "take into account the economic, social and environmental costs and benefits of the use of any pesticide," one would expect that pesticide i would be cancelled for use on crop j if the value of the vector of risks associated with use, $\mathbf{R}_{ij}$, exceeded the weighted sum of benefits of use, $\mathbf{B}_{ij}$. Treating unmeasured components of risks and benefits, u_{ij}, as random, the probability that pesticide i is cancelled for use on crop j is

$$P(\text{Cancel}_{ij}) = P(\alpha_1\mathbf{R}_{ij} + \alpha_2\mathbf{B}_{ij} + u_{ij} \geq 0), \tag{1}$$

where α_1 and α_2 are the vectors of policy weights attached to risks and benefits, respectively.

Special interest groups enter the model by augmenting the vectors of risks and benefits considered by EPA, or by altering the policy weights attached to risks and benefits. Suppose for example that $\mathbf{X}$ is a vector of variables indicating intervention in the policymaking process by each of several special interest groups. Then the general model becomes

$$P(\text{Cancel}_{ij}) = P(\alpha_1\mathbf{R}_{ij} + \alpha_2\mathbf{B}_{ij} + \delta_1\mathbf{X}_{ij} + u_{ij} \geq 0). \tag{2}$$

An alternative to equation (2) frequently proposed by researchers in the risk assessment area is that risks and benefits are balanced only for intermediate risk levels but not when risks are very high or very low. This so-called "bright line" theory of risk regulation hypothesizes that a health or safety regulation will always be undertaken if the risk to the maximally exposed individual, R, exceeds some risk threshold, R_{max}, and will never be adopted if the risk to the maximally exposed individual falls below some critical level, R_{min}. Between these thresholds, the theory holds, the regulation will be adopted if the risks outweigh the benefits. Formally,

$$P(\text{Cancel}) = 1 \qquad \text{if } R \geq R_{max},$$

$$P(\text{Cancel}) = \text{eq. (2)} \qquad \text{if } R_{max} > R > R_{min}, \tag{3}$$

$$P(\text{Cancel}) = 0 \qquad \text{if } R \leq R_{min}.$$

In adapting this theory to the case of pesticide cancellations, we note that there are three groups of individuals whose health EPA is supposed to protect: consumers of food products, pesticide applicators, and those who mix and load the pesticides. Because of differences in the magnitude and degree of voluntariness of the risks facing these three groups, it is plausible that EPA, if it follows (3), sets different risk thresholds for each of the three groups, and will cancel a registered use if the risk to the maximally exposed individual in any of the three groups exceeds the relevant threshold. On the other hand, for the substance to be judged "safe" it must fall below the R_{min} for each group. Because equation (2) is nested in (3) we can statistically test one hypothesis against another.

If sufficient information were available, the models in equations (2) and (3) could be estimated both for the proposed decision to cancel a pesticide registration and for the final decision. Unfortunately, lack of information about the role of intervenors prior to the public comment period makes estimation of either model impossible for the proposed decision. Although it is well known that EPA meets with interested parties throughout the Special Review process, it is only since 1985 that information about such meetings was required to be made public. Since we know only about interventions that occurred after the preliminary decision was made, our analysis is confined to explaining whether or not a pesticide was cancelled in EPA's Notice of Final Determination.

Variable Selection and Treatment of Missing Values

To explain EPA's final decision we have gathered data on the cancer and other health risks, and the benefits, associated with each food crop for which the 19 pesticides listed in table 1 were registered before entering Special Review. We have also attempted to measure the participation of interest groups. The variables for which sufficient observations are available are listed in table 2 and are described below.

Risk Variables

The (individual) risk associated with use of pesticide i on crop j is correctly computed as the difference between the lifetime cancer risk associated with pesticide i and the risk associated with the pesticide that will replace it if it is cancelled. EPA's published risk estimates, however, measure the risk of pesticide i as the incremental lifetime cancer risk associated with that pesticide, as though the alternative to using pesticide i were riskless. In this and in other instances cited below, we used EPA's published figures even if they do not measure the theoretically correct construct, because it is these figures that were available to decisionmakers.

In addition to measuring the maximum individual risk to applicators, mixer/loaders, and consumers of food products, we would like to measure the number of expected deaths associated with pesticide i on crop j. For cancer risks, however, data on the size of the exposed population is seldom reported. This poses little problem for measuring dietary risks, which are usually based on total U.S. food consumption and, hence, the total U.S. population, but is problematic for occupational exposures. Because the size of the exposed population is unavailable, the risk variables in our model represent risks to the maximally exposed individual, which, in practice, is the average applicator or mixer loader. These can be scaled to represent the expected number of deaths caused by the pesticide annually, as long as it is assumed that the size of the exposed population is constant across all observations.[7]

---

7. We cannot, however, distinguish the individual contributions of size of exposed population and individual risk to the regulatory decision.

TABLE 2.   Means and Standard Deviations of Variables Used in Model

Variable Name	Uses That Were Banned			Uses That Were Not Banned		
	Number of Observations	Mean	Standard Deviation	Number of Observations	Mean	Standard Deviation
Whether cancelled	96	1.0	.0	149	.0	.0
Dietary risk*	78	9.6E-4	3.5E-3	94	4.2E-6	1.4E-5
Applicator risk	63	1.2E-2	2.1E-2	66	1.5E-4	7.3E-4
Mixer risk	42	2.2E-4	8.8E-4	35	1.2E-5	9.9E-6
Producer benefits†	86	2.873	7.637	81	15.685	41.453
Whether yield loss	96	.240	.429	149	.530	.501
Reproductive effects	96	.458	.501	149	.376	.486
Danger to marine life	96	.583	.495	149	.470	.501
Environmental groups comment	96	.729	.447	146	.329	.471
Academics comment	96	.104	.307	146	.390	.490
Growers comment	96	.042	.201	146	.144	.352

*All risks are risks of cancer based on a lifetime of exposure to the pesticide.
†Millions of 1986 dollars.

Noncancer health risks and ecological risks are inherently difficult to measure. Noncancer health effects are measured by a dummy variable indicating that the pesticide exhibits adverse reproductive effects. Ecological risks are measured using a dummy variable that indicates whether a substance is harmful to marine life.

Benefit Data

The only measure of benefits to consumers and producers that is consistently provided in the risk-benefit studies EPA conducts is the losses that producers would sustain in the first year after cancellation of the pesticide. These are measured as the increased control costs from switching to a substitute pesticide and the value of any yield losses. If yield losses are large enough to raise the price of the product, losses to producers are reduced by the resulting increase in revenues. Because losses to consumers are seldom quantified in the background documents, we rely exclusively on first-year losses to producers.

Even the latter, however, are not available for all pesticides and all crops. It should be emphasized that, while pesticide manufacturers are responsible for data on health risks, EPA must bear the cost of calculating the benefits of pesticide use. If information on number of acres treated and on input and output prices are not available from other sources (e.g., USDA), budgetary limitations make it unlikely that benefits will be calculated. Even when such data are available, uncertainty about yield losses makes the cost of pesticide cancellation hard to quantify. When producer benefits are not measured in dollars, we use a dummy variable to indicate whether cancellation of the pesticide would result in yield losses to producers.

Political Variables

Quantifying the participation of special interest groups is a difficult task. Because the only information publicly available is whether comments were entered following the proposed decision, we use dummy variables that indicate whether such comments were made by at least one member of each interest group. The interest groups we distinguish include environmental groups, who commented on 49% of all decisions, grower organizations, who commented on 10% of all decisions, and academics, who commented on 28% of all decisions. The results below suggest that academics most often commented on behalf of growers or manufacturers.

One group whose influence we are unable to measure is pesticide manufacturers. Because these manufacturers comment on virtually every decision, the use of a registrant dummy is unproductive. Ideally, one would like to measure the financial stake that manufacturers have in individual pesticides, but such information is proprietary.

To capture the effect of one particularly controversial political administration, a dummy variable is included for the years in which Anne Burford was Administrator of EPA.[8]

Treatment of Missing Values

One problem with the data is the large number of missing values, especially for cancer risks and producer benefits (see table 2). In the case of cancer risks, data may be missing either because an estimate of dietary or occupational exposure is unavailable for a particular crop, or because toxicological data are not deemed sufficiently reliable to estimate a dose-response relationship. Although both situations occur in the data, it is the latter that accounts for the majority of missing observations.

A similar problem occurs with producer benefits from pesticide use. Because EPA does not have the budget to launch a primary data collection

---

8. The fact that we have only 19 active ingredients prevents more extensive use of political dummy variables in the probit model. If, for example, a dummy variable were added for each political administration, the Carter Administration dummy would explain perfectly all decisions on DBCP, the only pesticide to complete the Special Review process during that administration.

effort, lack of information from secondary sources about acres of the crop treated or about input and output prices makes it likely that benefit data will not be quantified.

We handle missing data problems by defining an indicator variable M_{ij} ($=1$ if data are missing) and multiplying the variable of interest (e.g., dietary cancer risk) by $1 - M_{ij}$. The missing data indicator also appears as an independent variable. The coefficient of the risk variable thus represents the effects of dietary cancer risk conditional on such information being available.

IV. The Determinants of Pesticide Decisions, 1975–89

Estimates of equations (2) and (3) appear in table 3 for three different sets of variables: (i) risk and benefit variables only; (ii) risk and benefit variables augmented by commenter dummies; (iii) all of the preceding variables augmented by a dummy variable indicating the Burford administration at EPA.[9]

Are Risks and Benefits Balanced?

Table 3 indicates that EPA does balance risks and benefits in deciding whether or not to ban a pesticide. Indeed, for each set of variables, the "bright line" theory of risk regulation, which asserts that risks and benefits are not balanced for very low or very high risk levels, can be rejected in favor of a simple probit model. In examining the so-called bright line theory (cols. 4–6), we note that there are no risk levels below which all pesticide uses were allowed. For example, some uses of captan were banned even though incremental risks to applicators were 10^{-9} and incremental dietary risks were 10^{-12}, presumably because benefits from captan usage were very small.[10] The maximum acceptable risk levels in our data (levels above which all uses were banned) differ somewhat from the 10^{-4} cutoff often emphasized in the risk management literature (Travis and Hattemer-Frey (1988)). For instance, the maximum acceptable risk level is highest for applicators (1.1×10^{-2}) but somewhat closer to conventional levels for mixers (3.1×10^{-5}) and for dietary risks (1.7×10^{-4}).

Because the bright line models were estimated by maximum likelihood

---

9. Equations (2) and (3) were estimated by maximum likelihood methods, assuming that $u_{ij} \sim IN(0, \sigma^2)$ for all i and j. Details on the estimation of the switch points in equation (3) appear in the Appendix. The three observations on ethalfluralin were dropped from the analysis since no information on public comments was available for these decisions.

10. EPA banned the use of captan on 44 fruits and vegetables. In each case, the benefits of captan use were estimated to be negligible. Average dietary risk was $\leq 10^{-6}$ in all cases and $\leq 10^{-9}$ in 28 cases. Risks to mixer/loaders and applicators were 10^{-6} in about half the cases and 10^{-5} in the other half.

TABLE 3. Probability of Cancellation Equations

	Continuous Model			Bright Line Model		
	(1)	(2)	(3)	(4)	(5)	(6)
Intercept	-.050	-.822	-1.824	.589	-.852	-1.391
	(.344)	(1.091)	(.785)	(.451)	(1.167)	(1.035)
Diet risk per million persons	.003	.009	.012	-.018	-.027	-.030
	(.006)	(.006)	(.006)*	(.030)	(.039)	(.036)
Diet risk missing	-.864	-.626	-.775	-1.020	-.718	-.821
	(.386)*	(.565)	(.540)	(.404)*	(.575)	(.556)
Applicator risk per million persons	4.4E-4	8.2E-4	6.7E-4	4.2E-4	7.8E-4	6.2E-4
	(2.2E-4)*	(3.0E-4)*	(2.7E-4)*	(2.3E-4)	(3.2E-4)*	(2.8E-4)*
Applicator risk missing	.554	-.836	-.529	.471	-1.09	-.869
	(.361)	(.672)	(.630)	(.364)	(.690)	(.667)
Mixer risk per million persons	.005	8.2E-4	1.5E-4	-.053	-.007	-.022
	(.008)	(2.5E-2)	(1.3E-2)	(.027)	(.042)	(.037)
Mixer risk missing	-.860	.681	.540	-1.374	.881	.644
	(.414)*	(.726)	(.683)	(.482)*	(.823)	(.785)
Producer benefits[†]	-.048	-.074	-.066	-.050	-.075	-.069
	(.018)*	(.028)*	(.025)*	(.019)*	(.0282)*	(.025)*
Producer benefits missing × yield loss	-1.984	-2.420	-2.413	-2.043	-2.46	-2.47
	(.361)*	(.454)*	(.446)*	(.379)*	(.460)*	(.458)*
Producer benefits missing × no yield loss	-1.797	-.296	-.934	-1.794	-.273	-.943
	(.446)*	(.889)	(.789)	(.447)*	(1.00)	(.862)
Reproductive effects	.530	.908	1.026	.843	1.03	1.13
	(.324)	(.532)	(.518)*	(.360)*	(.552)	(.537)*

(Continued)

TABLE 3.—*Continued*

	Continuous Model				Bright Line Model	
	(1)	(2)	(3)	(4)	(5)	(6)
Danger to marine life	.782	−.893	−.283	.617	−.990	−.593
	(.281)*	(.861)	(.701)	(.293)*	(.869)	(.769)
Burford years	...	−1.621	...	...	−1.38	...
		(1.112)			(1.22)	
Academics comment	...	−1.807	−1.333	...	−1.91	−1.58
		(.902)*	(.753)		(.914)*	(.819)
Growers comment	...	−2.017	−1.829	...	−2.42	−2.21
		(.874)*	(.762)*		(1.02)*	(.871)*
Environmental groups comment	...	3.070	3.398	...	3.27	3.48
		(.617)*	(.604)*		(.67)*	(.667)*
R_{max} diet	...	...	...	1.7E-4	1.7E-4	1.7E-4
R_{max} applicator	...	...	...	1.1E-2	1.1E-2	1.1E-2
R_{max} mixer/loader	...	...	...	3.1E-5	3.1E-5	3.1E-5
Log likelihood	−87.0	−44.9	−46.0	−83.4	−42.8	−43.4
Percentage of decisions correctly predicted	86.0	95.0	95.0	84.0	94.0	94.0

Note: Standard errors appear in parentheses below coefficients.
*Significant at the .05 level, two-tailed test.
†Millions of dollars.

techniques (see Appendix), likelihood ratio tests were performed to test the null hypothesis that bright lines do not exist (i.e., that (2) is the correct model) against the alternative that they do. In all three cases the null hypothesis cannot be rejected at conventional levels. We therefore focus our discussion on the simple probit results (columns (1)–(3) of table 3).

Given that EPA does weigh risks and benefits, what weight does it place on risks to different populations? In considering cancer risks, EPA clearly places most weight on risks to applicators. This variable is significant in all probit equations, and the ratio of its coefficient (suitably scaled) to that of producer benefits implies a value per statistical cancer case avoided of roughly $35 million (1986 dollars).[11] By contrast, risks to mixers are insignificant in determining the probability of cancellation, and dietary risks are significant at conventional levels only in column 3. The value per cancer case avoided implied by this coefficient, however, is only $60,600.

It is interesting to speculate on the reasons for these results. One reason for placing so much weight on reducing risks to applicators is that applicators constitute an identifiable population who face large risks. It is certainly plausible that equivalent risk reductions (in terms of numbers of cancer cases) are valued more highly when the level of individual risk is high (as it is for applicators) than when it is low (as it is for consumers). It may also be the case that decision makers discount risk estimates based on dietary exposure, which are widely known to be upward biased, relative to risk estimates for applicators, which are based on more accurate estimates of exposure.[12]

As far as other risks are concerned, the presence of adverse reproductive effects increases the probability of cancellation, although this effect is only marginally significant in columns (1) and (2). Danger to marine life, however, raises the probability of cancellation only in column (1). One reason for this may be the presence of comments by environmental groups in columns (2) and (3). As will be shown below, environmental groups are more likely to comment when a pesticide poses danger to marine life; hence, the environmental group dummy in equations (2) and (3) may be capturing some of the effects of this risk variable.

Producer benefits significantly lower the probability of cancellation. A $1 million increase in producer benefits lowers the probability of cancellation (with all variables at median values) between 0.7 and 1.1 percentage points.

---

11. The exact figures for columns (1), (2) and (3) of table 3 are, respectively, $32.1 (14.2), $38.8 (21.2), $35.5 (19.5) million (standard errors in parentheses). (These calculations are explained in detail in the Appendix.) It is interesting to note that intervenors do not significantly change the value per cancer case avoided.

12. Estimates of dietary cancer risk are usually based on the assumptions that pesticide residues are present at the maximum levels allowed by law and that the pesticide is used on all acres of the crop in question.

Even when producer benefits are not quantified, merely knowing that yield losses would occur if the pesticide were banned significantly reduces the probability of cancellation.

How Important Are Political Interests in the Regulatory Process?

The dramatic increase in the log of the likelihood function when interest group variables are added to the model attests to the importance of intervenors in the regulatory process. Participation by environmental groups dramatically increases the probability of cancellation, while participation by grower organizations and academics reduces the probability of cancellation. From this we infer that most academics are commenting on behalf of growers or of registrants.

While having the expected sign, the dummy variable indicating the Burford period at EPA is not significant at conventional levels, although it alters somewhat the magnitude of the coefficients on the interest group variables. This fact prompts us to examine whether Ms. Burford may have exerted influence indirectly by discouraging comments from environmental groups and encouraging comments from growers. To investigate this issue we estimated separate probit models to explain the participation of environmental groups and grower organizations (see table 4). The Burford regime appears to have influenced environmental groups, as none of them bothered to enter comments in the public docket during her administration. (The Burford dummy does not appear in the environmental group equation because it would have a coefficient of $-\infty$.) By contrast, she appears to have increased the probability that grower organizations would comment. This may reflect the fact that environmental groups felt it futile to intervene during Burford's tenure, whereas grower organizations expected a more sympathetic hearing.

The results of table 4 also shed light on an issue raised earlier. To some extent, participation by interest groups in the regulatory process is motivated by the risks and benefits of pesticide use that an unbiased "social planner" would consider. Environmental groups, for example, are more likely to comment on pesticides that pose a danger to marine life, and grower organizations are more likely to comment the larger are benefits to them from pesticide use. Because some of the factors that EPA is required to consider under FIFRA may be captured by intervenor dummies, one should not be surprised if, as in columns (2) and (3), variables such as Danger to Marine Life and Producer Benefits become less significant than they appear in column (1).

Finally, something should be said about the effect of the proposed decision to cancel a pesticide on the likelihood that interest groups comment. While it is certainly plausible that a proposed decision to cancel a pesticide

Table 4. Probability of Commenting Equations

	Environmental Groups (1)	Grower Organizations (2)
Intercept	−1.073	−2.634
	(.211)	(.381)
Reproductive effects	.077	
	(.218)	
Danger to marine life	.471	
	(.201)*	
Proposed decision = cancel	1.615	1.550
	(.222)*	(.365)*
Burford years		.973
		(.296)*
Producer benefits†		.011
		(.004)*
Producer benefits missing		−.196
× yield loss		(.345)
Producer benefits missing		−.334
× no yield loss		(.547)
Log likelihood	−115.0	−62.3
Percentage of decisions		
correctly predicted	93.0	90.0

Note: Standard errors appear in parentheses below coefficients.
*Significant at the .05 level, two-tailed test.
†Millions of dollars.

increases the chances that growers will comment, it is puzzling that a proposed decision to cancel increases the chances that environmental groups comment. It is, after all, environmental groups who usually oppose pesticide use. The positive sign here may, however, reflect reverse causality: by exerting influence before as well as during the public comment period, environmental groups may actually increase the chances of a proposed cancellation.

V. Conclusions

We suggested in the introduction to this paper that our findings would both comfort and concern those interested in environmental regulation. With respect to the former, it appears that EPA is indeed capable of making the kind of balancing decisions that economists presumably support and that FIFRA clearly requires. Our results convincingly demonstrate that the existence of risks to human health and/or the environment increases the likelihood that a particular pesticide use will be cancelled by EPA; at the same time, the larger the economic benefits associated with a particular use, the lower the likelihood of cancellation.

On the other hand, our results also provide some cause for concern. For instance, we find that the value of a statistical life implicit in the 242 regulatory decisions we consider is $35 million for applicators but only $60,000 for consumers of pesticide residues on food. Why is the EPA apparently willing to spend nearly 600 times as much to protect those who apply pesticides as those whose exposures come through food residues? Two explanations seem likely. First, although they are many fewer in number, each applicator faces a much larger individual risk than a typical consumer—on average about 15 times larger. EPA may be especially concerned about allowing larger individual risks. Second, because they are smaller in number, applicators are more identifiable than the more than 200 million consumers of food in the U.S. As with the proverbial baby in the well, society stands willing to spend much more to save the lives of identifiable victims than mere "statistical lives," and this may be reflected in our findings.

There are other aspects of the pesticide regulatory process that provide some cause for concern. First, although hardly unique to pesticide regulation, the procedures used to assess risks to all parties are almost sure to lead to upwardly biased estimates. To take but one example from the decisions analyzed here, risks to applicators and consumers are predicated on the assumption that no other active ingredient will be substituted for one banned in a particular use. Since such substitutions are the rule rather than the exception, however, a more accurate measure of risk reductions would reflect the *differential* riskiness of the two substances. (It is conceivable, in fact, that a more hazardous—yet to date untested—ingredient could be substituted for one whose use was discontinued by EPA.) This suggests, incidentally, that groups of active ingredients be considered together in the regulatory process. This would encourage more accurate estimation of both risks and benefits since it would make clear those situations where simple substitutions are no longer possible. Finally, EPA should be given the resources to make more accurate estimates of the benefits of pesticide usage. It is simply not sufficient to calculate losses to growers and call this the "cost" of restricting a particular pesticide use. More sophisticated measures, which include foregone consumers' surpluses, must become a standard part of FIFRA regulation.

It is less clear how one should view our findings concerning the political variables we examined. Clearly, intervention in the regulatory process—by both business and environmental groups—affects the likelihood of pesticide use restrictions. All other things being equal, interventions by environmental groups have about twice the impact on the likelihood of cancellation than those by growers (although the combined effect of growers and academic commenters, who weigh in against cancellations, outweighs that of environmentalists). Moreover, Anne Burford's short and controversial tenure at EPA is seen to have had a negative effect on the likelihood of pesticide cancellations. To those who view pesticide or other similar regulation as the proper province of scientists, engineers, and economists alone, these findings may be discouraging. On the other hand, those taking the view that regulation—like government taxation or spending—is inherently a political act, may find it encouraging that affected parties not only participate actively in the regulatory process but do so quite effectively.

APPENDIX

Maximum Likelihood Estimation of Bright Lines

Our risk levels R_{max} have been chosen by ordering observations from most to least risky and finding the lowest risk level in each category (diet risk, applicator risk, mixer risk) above which all uses were cancelled.

It can be argued that this is equivalent to picking R_{max} to maximize a likelihood function for which the individual terms are:

$$
\begin{aligned}
P(\text{cancel}) &= 1 & &\text{if } R \geq R_{max}, \\
P(\text{cancel}) &= \Phi(\alpha_1 R + \alpha_2 B) & &\text{if } R < R_{max}, \\
P(\text{don't cancel}) &= 0 & &\text{if } R \geq R_{max}, \\
P(\text{don't cancel}) &= 1 - \Phi(\alpha_1 R + \alpha_2 B) & &\text{if } R < R_{max},
\end{aligned}
\tag{A1}
$$

where Φ is the standard normal distribution function.

The argument that our procedure maximizes the likelihood function is as follows: If one were to raise R_{max}, this would take observations that were cancelled and now contribute a "1" to likelihood function and reduce their contribution to $\Phi(\alpha_1 R + \alpha_2 B)$ ≤ 1, thus lowering the value of the likelihood function. If one were to lower R_{max}, observations that were not cancelled would now be above the R_{max}. From equation (A1) above, the contribution of these observations to the likelihood function would fall to zero from $1 - \Phi(\alpha_1 R + \alpha_2 B) \geq 0$, thus lowering the value of the likelihood function. Our procedure thus maximizes the likelihood function.

We can, therefore, view the threshold model as consisting of 242 observations and $k + 3$ parameters, where k is the number of parameters estimated in the continuous model. A test of the threshold model can be conducted by comparing $2(\ln L(\text{threshold}) - \ln L(\text{continuous}))$ with the critical value of the χ^2 distribution with 3 degrees of freedom.

Calculation of Implied Value per Cancer Case Avoided

Equation (1) in the text indicates that pesticide i will be cancelled for use on crop j if the risks associated with the pesticide, R_{ij}, plus other considerations, u_{ij}, outweigh the benefits of use, B_{ij},

$$\alpha_1 R_{ij} + \alpha_2 B_{ij} + u_{ij} > 0 \ .$$

Equivalently, the pesticide will be banned if the value of the risks outweighs the dollar value of the benefits,

$$-\frac{\alpha_1}{\alpha_2} R_{ij} - \frac{u_{ij}}{\alpha_2} \geq B_{ij}. \tag{A2}$$

If R_{ij} is the number of cancer cases avoided by banning the pesticide for a year and B_{ij} is the annual value of benefits, then $-\alpha_1/\alpha_2$ is the value per cancer case avoided.

In estimating α_1 and α_2, R_{ij} has been replaced by N_{ij}, the number of cancer cases avoided per million exposed persons, based on a lifetime (T years) of exposure. The relationship between R_{ij} and N_{ij} is thus given by

$$R_{ij} = \frac{N_{ij}}{T \times 10^6} \times \text{number of persons exposed}, \tag{A3}$$

where the first term on the right-hand side of (A3) is the risk to a single person from a year of exposure to the pesticide. Equation (A3) implies that the coefficient of N_{ij} must be multiplied by $T \times 10^6$ and divided by the number of persons exposed to equal α_1.

To calculate the value per applicator cancer case avoided, the coefficient of applicator risk (N_{ij}) must be divided by the number of applicators exposed and multiplied by 35×10^6. (A lifetime of exposure for an applicator is assumed to be 35 years.) The resulting estimate of α_1 must then be divided by minus the coefficient of producer benefits. To illustrate the calculation, we use the coefficients in the third column of Table 4 and assume an exposed applicator population of 10,000. This implies a value per applicator cancer case avoided of \$35.53 million (1986 dollars):

$$-\frac{\alpha_1}{\alpha_2} = \frac{6.7 \times 10^{-4}}{.066} \frac{35 \times 10^6}{10,000} = 35.53.$$

In calculating the value per cancer case avoided associated with dietary risks, $T = 70$ and $N = 2.1 \times 10^8$, i.e., the U.S. population during the period of study.

REFERENCES

Bosso, Christopher J. 1987. *Pesticides and Politics: The Life Cycle of a Public Issue.* Pittsburgh: University of Pittsburgh Press.

Crandall, Robert W. 1983. *Controlling Industrial Pollution: The Economics and Politics of Clean Air.* Washington, D.C.: The Brookings Institution.

Hird, John A. 1990. "Superfund Expenditures and Cleanup Priorities: Distributive Politics or the Public Interest?" *Journal of Policy Analysis and Management* 9:455–83.

McFadden, Daniel. 1976. "The Revealed Preferences of a Government Bureaucracy: Empirical Evidence." *Bell Journal of Economics and Management Science* 7:55–72.

———. 1975. "The Revealed Preferences of a Government Bureaucracy: Theory." *Bell Journal of Economics and Management Science* 6:401–16.

Magat, Wesley; J. Krupnick; and Winston Harrington. 1986. *Rules in the Making: A Statistical Analysis of Regulatory Agency Behavior.* Baltimore: Johns Hopkins University Press.

Milvy, P. 1986. "A General Guideline for Management of Risk from Carcinogens." *Risk Analysis* 6:69–79.

Morrall, John F. III. 1986. "A Review of the Record." *Regulation* 10:25–44.

Nichols, Albert L., and Richard J. Zeckhauser. 1986. "The Perils of Prudence: How Conservative Risk Assessment Distorts Regulation." *Regulation* 10:13–24.

Pashigian, Peter. 1985. "Environmental Regulation: Whose Self-Interests Are Being Protected?" *Economic Inquiry* 23:551–84.

Peltzman, Sam. 1976. "Toward a More General Theory of Regulation." *Journal of Law and Economics* 19:211–40.

Stigler, George J. 1971. "The Theory of Economic Regulation." *Bell Journal of Economics and Management Science* 2:3–21.

Thomas, Lacy Glenn. 1988. "Revealed Bureaucratic Preference: Priorities of the Consumer Product Safety Commission." *Rand Journal of Economics* 19:102–13.

Travis, Curtis C., and Holly A. Hattemer-Frey. 1988. "Determining an Acceptable Level of Risk." *Environmental Science and Technology* 22:875–76.

Travis, Curtis C., Samantha A. Richter, Edmund A. C. Crouch, Richard Wilson, and Ernest D. Klema. 1987. "Cancer Risk Management." *Environmental Science and Technology* 21:415–20.

U.S. Environmental Protection Agency. 1989. "Report on the Status of Chemicals in the Special Review Program and Registration Standards in the Reregistration Program." Washington, D.C.: Office of Pesticide Programs.

Weingast, Barry R., and Mark J. Moran. 1983. "Bureaucratic Discretion or Congressional Control? Regulatory Policymaking by the Federal Trade Commission." *Journal of Political Economy* 91:765–800.

White, Lawrence J. 1981. *Reforming Regulation: Process and Problems.* Englewood Cliffs, NJ: Prentice-Hall.

Yandle, Bruce. 1989. *The Political Limits of Environmental Regulation: Tracking the Unicorn.* Westport, CT: Quorum Books.

CHAPTER 7

Planning versus Reality: Political and Scientific Determinants of Outer Continental Shelf Lease Sales

Porter Hoagland and Scott Farrow

Introduction

In 1987, a government agency planned 38 federal actions—the sale of leases for offshore oil and gas lands—to take place over a period of five years in 22 locations. At the end of that period, however, the agency had held only 15 sales in only five locations. Whether the difference between the planned action and the actual outcome constituted a success or failure of environmental policy depends on the perspective of the various interest groups involved.

Political interest groups, bureaucrats, and scientists all have a vested interest in understanding what determines the environmental policy decisions to which they contribute and in which they have a stake. Beyond these players, individual citizens and businesses also have a stake in how environmental policies are determined, given that it has been estimated that environmental policies (narrowly defined to exclude natural resource policies) cost more than $100 billion per year and are rising (EPA 1990; Hazilla and Kopp 1989). In addition, policies such as wetland regulations, oil and gas leasing in the Arctic National Wildlife Refuge (ANWR) and on the Outer Continental Shelf (OCS), and threatened or endangered species decisions highlight the political and economic significance of natural resource issues in more broadly defined environmental policy.

This chapter investigates the political, scientific, and organizational factors that contributed to a multibillion dollar decision about the disposition of natural resources—the number of lease sales to be permitted on the federal ocean lands, known as the Outer Continental Shelf—between 1987 and 1992. In the case under consideration, only 40 percent of the planned sales took place and then in only 25 percent of the designated locations.

This research was funded in part through the Center for Business and Government, Harvard University and by The Woods Hole Oceanographic Institution.

What determines the number and location of planned and actual sales? Adherents of rational government planning might well claim that the answer is natural and social scientific data analysis. Those who believe in the primacy of political factors would suggest that political influence is the overriding factor. Devotees of organizational theory, on the other hand, would look to the procedures of the relevant institutions. A public choice approach can help us structure models from each of these fields. What follows is an examination of rational government planning within the broader context of the influence of interest groups. We begin with the theoretical and empirical literature, go on to discuss the institutional setting of our case study, and conclude with a statistical analysis of the planned and actual decisions.

Theoretical Structure and Prior Results

Several lines of inquiry by economists have merged into the public choice literature. For our purposes, the most relevant approach deals with the economic behavior of interest groups in settings of legislative policy making and bureaucratic decision making. This literature grew out of the theoretical work of Buchanan and Tullock (1962), Olson (1965), Stigler (1971), Becker (1983), and McFadden (1976) and evolved into the rent-seeking literature surveyed in Mueller 1989; Buchanan, Tollison, and Tullock 1980; and Rowley, Tollison, and Tullock 1988.

In these nonmarket settings, decision making is often made either by a voting process or by a legally constituted centralized bureaucratic authority—a specified person or group. Interest groups, broadly defined, compete for the capture of rents which can distort both the production frontier of the economy and the distribution of income.

Outer Continental Shelf planning and implementation can be described as an example of a single decision maker, the secretary of the interior, maximizing his or her independent utility function.[1] But, as the burgeoning literature on interest group influence attests, such a characterization can be too myopic. In fact, the legal authorization for decision making found in the OCS Lands Act and its amendments merely sets procedural constraints on a multiparty negotiation. As we shall see, if politically powerful interest groups face potentially unfavorable outcomes, the bureaucratic process can be circum-

---

1. Proponents of the 1978 OCSLA Amendments hoped that rational and scientific management would result from the "progressivist" FYLP process. Heintz (1988, 228) explains that the pressures to reduce oil shortages (as reflected in gas lines), inflation, and unemployment were so great during the Carter administration that "[i]ronically, central planning worked against the anti-leasing coalition which had for so long advocated it. The [1979 Department of Energy] analysis of OCS production goals strengthened the arguments in favor of more rapid leasing, a result quite opposite of that intended by the anti-leasing coalition."

vented. This circumvention creates new constraints or alters the broader political calculus of the decision maker.

Here we develop a suite of models which first structure the polar effects of either political or scientific influences on bureaucratic decision making. A mixed model of effects naturally follows. Our data set allows the application of the models to both planned and actual decisions so that we can investigate changes in the statistical determinants of the decision.

Many political and bureaucratic models involve discrete or ranked choices (e.g., McFadden 1976; Cropper et al. 1992). Our concern is with a cardinal choice, since the decision centers on the number of oil and gas lease sales to be held in various areas. While formally this is a problem in integer programming, we develop the basic framework in a manner similar to that of Magat, Krupnick, and Harrington (1986) in their analysis of the stringency of environmental standards.

We begin with a purely political model in which the secretary considers the political impact of holding a specified number of offshore oil and gas sales in area i, denoted S_i. The political impact need not be in the form of votes for the existing administration or for congressional members associated with oversight of the agency. Instead, the impact may represent delays in carrying out policy, changes in budget, or any other situation in which bargaining among groups with power may affect the preferences of the secretary.

Define a net political impact objective function, P, which is the difference between the secretary's perceptions of the political benefits of holding sales S_i, $f(S_i)$, and the perceived political costs, $h(S_i)$:

$$P = f(S_i) - h(S_i).$$

One can naturally associate certain interest groups with such respective functions as the oil and gas industry and environmental groups. This structure of the problem (and the equating of marginal values for an optimum) suggests that factors which affect the secretary's perceptions of the political benefits and costs can alter the decision. Constraints on the problem can also become important. A substantial part of the case study involves the imposition of congressional and presidential moratoria which limit some specific S_i to be equal to zero.

An alternative rational planning model suggests that scientific information—information that is not related to the bargaining power of interest groups—represents the determinants of the number of sales. Consider a net scientific impact objective function which quantifies the benefits and costs in dollar terms as much as possible but which also leaves some scientific information to be qualitatively factored into the decision. Define $B(S_i)$, $C(S_i)$ and $O(S_i)$ to be respectively the monetary benefits, monetary costs, and the other non-monetized scientific measures of impact, here related to costs.[2] The net scientific function becomes

$$E = B(S_i) - C(S_i) - O(S_i).$$

This formulation allows the possibility of a social welfare maximizer to equate the marginal social benefits with the marginal social costs.

It is natural as well to consider a mixed model in which the decision maker is neither a purely political player nor a purely scientific planner. In fact, the usual (though not required) exclusion of distributional impacts from benefit-cost analysis suggests that government decisions may quite appropriately mix decision-making criteria.

A linear combination of the models may be represented by

$$Y = d \ (\mathbf{E}) + (1 - d) \ (\mathbf{P}),$$

where Y is the decision and d is the weight given to the different models. Tests which we conduct investigate whether d equals 1 (only science matters) or whether d equals 0 (only politics matters).[3]

The organization making the decision can also appear in a minor but testable way in this process. Like Magat, Krupnick, and Harrington (1986), we focus on the source of the specific information elements presented to the decision maker. We hypothesize that the decision maker may pay particular attention to the data generated by his or her organization, in this case the Department of the Interior. These data may be in either the political or the scientific model. The comparison between planned and actual outcomes in the data allows tests for changes in the importance of individual data elements as well as groups of elements. This focus on the dynamics between planned and actual outcomes is seldom available in the literature. Two exceptions are Magat, Krupnick, and Harrington (1986) and Hamilton (1993; also chap. 9 in this volume). Magat, Krupnick, and Harrington evaluate the modification of provisions that take place during the process separating the rules proposed by the contractor and the determination of the final rules. They conclude that the process had a moderating effect on initial proposals that were either strong or weak. Hamilton (1993) studied planned and actual expansions in hazardous waste disposal facilities in two different time periods, the mid-1970s and

---

2. These impacts could be positive or negative. In the case to be studied they are negative impacts related to costs.

3. A topic for further research would estimate the value of d when a mixed model cannot be rejected. Such a model, perhaps estimable with nonlinear methods or inferred through coefficients of partial determination, could reveal an index of the importance of politics and science in a manner that is comparable across specific studies.

mid-1980s, and found that collective actions as measured by voter turnout become statistically significant in the more recent time period.

Other empirical studies can be grouped according to whether they studied planned outcomes (e.g., Farrow 1991; Mohai 1987; Thomas 1988) or actual outcomes (e.g., McFadden 1976; Cropper et al. 1992; Weingast and Moran 1983). The empirical literature to date has demonstrated a variety of idiosyncratic influences of both politics and science on bureaucratic decision making. This chapter provides new evidence on the dynamics of both politics and science in one agency while building toward a set of hypotheses to allow comparisons across studies of separate agencies.

Case Study: Institutional Setting

In the United States, the federal government manages the submerged oil and gas lands of the OCS seaward of state-managed submerged lands which generally extend three nautical miles from shore.[4] Since the passage of the Outer Continental Shelf Lands Act of 1953 (OCSLA), the federal government has received nearly $100 billion in royalties and up-front payments (bonuses) for leases.[5] In 1993, OCS production accounted for almost 25 percent of U.S. domestic production of natural gas and almost 12 percent of domestic oil production.

Over time, increasing OCS production coincided with increasing public interest in environmental issues and with highly publicized accidents that resulted in damage to the environment. Following the 1969 Santa Barbara oil spill, political opposition to the program grew rapidly. Congress began to examine the program more closely when the Nixon and Ford administrations attempted to expand leasing as a response to the OPEC oil embargo in the mid-1970s. In 1978, Congress amended OCSLA significantly. The result was a five-year planning process for the sale of leases. It included a requirement that the agency analyze the impact of alternative levels of leasing activity and locations, along with a legislative mandate to balance multiple interests in the planning process.[6] A relatively specific set of information requirements was

---

4. Offshore Texas and Florida state lands extend three marine leagues from the baseline. Economic rents from federal leases in a buffer zone that exists within a three-mile band adjacent to the state-federal boundary are shared with the adjacent coastal state.

5. OCSLA was substantially amended in 1978.

6. Specifically, OCSLA section 18(a)(3) (added during the 1978 amendments) requires the Interior secretary to "select the timing and location of leasing, to the maximum extent practicable, so as to obtain a proper balance between the potential for environmental damage, the potential for the discovery of oil and gas, and the potential for adverse impact on the coastal zone" (43 U.S.C. sec. 1344[a][3] [1993]). See Heintz (1988) for a case study of the OCS program from 1973 to 1984.

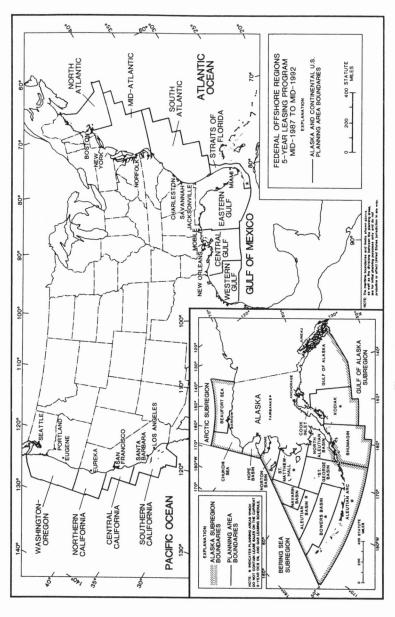

Fig. 1. OCS planning areas

established by OCSLA section 18(a)(2). This included collecting and analyzing information on economic, social, and environmental variables, on the interests of industry, and on other concerns (for an extended discussion, see Farrow et al. 1990).

The sales in our study were made under the 1987–92 plan, developed in the Reagan administration after James Watt had been replaced by Donald Hodel as interior secretary, and shortly after the dramatic 1986 decline in the price of oil. Implementation of the plan occurred primarily during the Bush administration. Relevant events during this period include establishment of a presidential task force to review OCS leasing, the war with Iraq, the *Exxon Valdez* spill, and the development of a National Energy Strategy.

The Five-Year Leasing Plan (FYLP) for these years determined the federally planned lease sales of OCS lands in areas off the shores of one or more states. Such sales are made through a sealed bid auction for tracts, which cannot exceed nine square miles in area. The current minimum bid for such a tract is $144,000; the maximum successful bid to date is $328 million. Exploration and possible development cannot proceed unless a company holds a lease. Consequently, the sale of such leases has become the focus for conflicts between those who favor development and those who oppose it. Figure 1 shows the planning areas for sales.

For example, the planning area of the Central Gulf of Mexico offshore from Louisiana and Mississippi represents the largest single offshore production area for oil and gas. In contrast, the lands offshore of California and Alaska fall into several different planning areas. The planned and actual number of sales in each area is the dependent variable in the models estimated in later sections.

Interest Groups

Numerous interest groups participated in the development of the FYLP. Congress has identified some of the interest groups identified in the OCSLA: the oil and gas industry, the fishing industry, and state governments. Various public interests have identified themselves during the comment and review process established under provisions of the Administrative Procedures Act for major policy actions (ACUS 1991). A brief survey of those interest groups will set the stage for the model of rent-seeking behavior offered below.

Identification of the prodevelopment interests of the oil and gas industry on the one hand and the antileasing interests of some environmental groups on the other is relatively straightforward. The industry is represented by the major as well as the smaller oil companies. The American Petroleum Institute, the Western States Petroleum Association, and the National Ocean Industry Association (offshore services) favor expanded OCS development. The leading environmental groups include the Natural Resources Defense Council (NRDC), the Sierra Club, and Greenpeace, who generally oppose expanded development (e.g., Holling 1990). The latter are joined by some states, as well as local environmental organizations. Coastal states have a stake in the lease sale decision because they may incur some costs (risks of oil spills, loss of tourism income, expanded demand for public services) or benefits (increased tax base, potential revenue sharing income) as a result of OCS development (CSO 1985). Some of the most politically active coastal states are California, Massachusetts, Florida, Louisiana, Alaska, and North Carolina. Local governments have similar concerns.[7]

Since approximately $20 million is spent annually to procure environmental studies in support of the offshore leasing program, marine environmental scientists have a stake in the program. In addition, the National Academy of Sciences is frequently asked to review the credibility of scientific information; its latest effort was commissioned by President Bush's task force and by the Department of the Interior (e.g., NRC 1989, 1992). Policy analysts (both within government and from the outside) regularly review the program for its revenue and production, the working of its auction process for granting individual leases, and for its social and environmental impacts (e.g., Hendricks and Porter 1992; Farrow 1987b; U.S. General Accounting Office 1985; Farrow and Rose 1992; Moody and Kruvant 1988).

The organization charged to develop the plan, the Minerals Management Service (MMS; an agency within the Interior Department), has its own organizational history, more rooted in a prodevelopment ethos than in the ethos of an environmental organization. At the top of the bureaucratic pyramid is a political appointee, the secretary of the interior. In developing the FYLP, the secretary receives information generated internally on what might be considered the variables of concern to all the interest groups: the net benefits of extracting oil and gas, the social costs, the regional distribution of costs and benefits, environmental sensitivity, and biological productivity. In addition, as required by OCSLA, the secretary receives input from outside the organization, including formally solicited indications of interest from industry, responses by state governors, and a range of input generated by the comment and review process of the Administrative Procedures Act.

The proposed FYLP consists of a set of staff-prepared sale options, supported by hundreds of pages of attached documentation linked to an environmental impact statement (U.S. Minerals Management Service 1987). The final FYLP represents the secretary's choice among the options presented for the number of sales in each planning area; his or her choice of modifications to the size of the area offered for sale; and possibly some additional program-

---

7. We will not focus on the specific question of the distribution of OCS revenues although that issue can be an element of citizen, state, and local government support or opposition.

matic decisions (e.g., whether or not to change the minimum bid per acre). Consequently, the planned resolution of the competing organizational, political, and scientific interests is officially the choice of the secretary, recognizing the complex process by which the document actually lands on his or her desk.[8] Other parties having stakes in the resolution, described in more detail later, are the president and the members of congress.

Case Study: Analysis

Farrow (1991) initially analyzed the economic and environmental science determinants of the secretary's decision for the planned sales in the 1987–92 FYLP. We extend this analysis to investigate more deeply the role of political interest groups and the mixed model previously discussed. The task is to identify quantifiable variables corresponding to the political, scientific, and organizational classes identified in the theoretical section.

Our scientific determinants of decision were highlighted in the FYLP as *Net Social Value* and an index of *Relative Marine Productivity and Environmental Sensitivity*. *Net Social Value* represents the Department of the Interior's estimates of the present value of the net benefits of energy extraction in each area, including the subtraction of social costs associated with oil spills. This variable measures the net monetized benefits for a scientific decision. *Relative Marine Productivity and Environmental Sensitivity* is an index that aggregates measures of phytoplankton productivity and sensitivity of marine habitats and biota in the planning area should an oil spill occur. This variable is taken to measure a nonmonetizable but potentially relevant scientific determinant of the decision.

Our political decision variables are a mixture of information highlighted in the FYLP as well as other information. The FYLP reports planning areas ranked by a survey of oil and gas producers. We take this variable, *Industry Rank,* to be a direct measure of the political interests of that group. The political interests of groups opposed to leasing are not as clearly represented in the FYLP.

We assembled a suite of variables that we believe measure in part the interest and power of the groups opposed to leasing. The variables we use for this group are: (1) the number of members of three conservation groups— Sierra Club, Greenpeace, and National Wildlife Federation and (2) a discretized index based on the League of Conservation Voters' score on seven key votes for the Clean Air Act Amendments of 1990. The variable *Environmental Sensitivity* is also assumed to be correlated with the degree of interest of groups opposed to leasing.

---

8. In fact, the document is prepared by MMS staff members, who receive advice and guidance from officials in Interior's Office of Program Analysis.

Clearly the distinction between scientific and political variables is not perfect. For instance, we can expect *Industry Rank* to be influenced by the major component of *Net Social Value,* the present value of extraction. Indeed, a moderate (below 0.6) correlation exists. Similarly, environmental preferences might be partially reflected in the oil spill cost component of the *Net Social Value* as well. However, political influence, like willingness to pay, requires both preferences and political power so that the broader set of variables represent an added dimension to that provided by the scientific information.

We have noted that there may also be a preference for relying on data generated within the organization. The variables *Net Social Value, Industry Rank,* and *Environmental Sensitivity* were all collected by the organization as part of its original planning process. The other data represent political information generated outside the organization.

A Poisson regression is used to estimate the statistical relationships. Since the sales numbers in each area are nonnegative integers, the Poisson provides a frequently used model for that kind of data. Conceptually the estimation problem is analogous to that of standard regression analysis. Parameters of the conditional mean of the Poisson are estimated assuming a functional form. As is typically done in Poisson analysis, the logarithm of the mean was modeled as a linear function in the level of the variables. This imposes, as desired, nonnegativity on the mean of the Poisson, while allowing for a natural interaction among the variables.[9]

Table 1 reports the results for the determinants of *planned* (ex ante) sales.

The political models are discussed first. Column 1 of the table reports a de minimis political model which includes only *Industry Rank.*[10] Taken alone, that rank is a significant determinant of planned sales. Column 2 adds the political variables for groups opposed to leasing. None of those variables is significant individually. We also conduct a likelihood ratio test for the joint significance of the added variables which continues to indicate insignificance for the opposition group at all standard levels. We conclude that a political model cannot reject industry rank as a determinant of the planned number of sales, although we do reject nonindustry interests as a determinant.

Columns 3 and 4 present a competing model of the decision based on scientific information. Taken alone, the *Net Social Value* is a significant

9. Parameter estimates were obtained using a maximum likelihood routine in the LIMDEP program.

10. The data, model specifications, and models used for estimating the parameters listed in table 1 were somewhat modified from those employed by Farrow (1991). The data changes include the addition of a cardinal regressor for environmental rank and the measures of political interest to groups opposed to leasing.

TABLE 1. Poisson Analysis of Planned Sales

	(1)	(2)	(3)	(4)	(5)	(6)
Constant	1.30	1.90	0.2	1.9	1.3	1.67
	(0.29)	(1.33)	(0.2)	(0.9)	(1.0)	(1.38)
	0.0	15.4	21.6	3.3	18.5	22.5
Net Social Value			4.3×10^{-5}		2.9×10^{-5}	0.27×10^{-4}
			(1.2×10^{-5})		(1.7×10^{-5})	(0.19×10^{-4})
			0.0		8.4	14.8
Industry Rank	-0.078	-0.0725			-3.5×10^{-2}	-0.39
	(0.028)	(0.028)			(3.6×10^{-2})	(0.37)
	0.5	1.1			33.0	29.4
Environmental Sensitivity		-0.3038		-5.9×10^{-3}	-2.7×10^{-3}	-0.003
		(0.0037)		(3.8×10^{-3})	(4.0×10^{-3})	(0.004)
		30.4		12.5	50.4	45.0
Conservation Members		-0.0005				-0.86×10^{-3}
		(0.002)				(0.21×10^{-2})
		81.4				68.0
League of Conservation Voters		0.126				-0.061
		(0.338)				(0.345)
		71.1				86.0
Log likelihood	-27.00	-27.13	-27.00	-31.00	-26.18	-26.07

Notes: Variable definitions are as follows: *Net Social Value* equals net present social value; *Industrial Rank* equals industry preferences, the larger number less preferred; *Environmental Sensitivity* equals relative marine productivity and environmental sensitivity, the higher number more sensitive; *Conservation Members* equals Sierra Club, Greenpeace, and National Wildlife Federation members; *League of Conservation Voters* is an index based on a discretized ranking by the League of Conservation Voters for senatorial votes on Clean Air Act Amendments of 1990.

Standard errors appear in parentheses.

P-values (in percent) under standard errors indicate significance level, for example, 5.0 means significant at 5 percent level.

determinant of the number of sales (col. 3) although the environmental science information is not a significant determinant of the number of sales, even when taken alone as in column 4. We conclude that a scientific planning model cannot reject the importance of economic measures, although an environmental science measure is rejected as a significant determinant of planned sales.

Mixed models of political and scientific determinants are reported in columns 5 and 6. The continuing statistical significance of *Net Social Value* and the insignificance of industry and environmental variables in the mixed models lead us to conclude that at the planning stage we cannot reject a rational planning model based on an economic measure. This conclusion is moderated somewhat by the effect noted by Farrow (1991) that the size of the coefficient of *Net Social Value* is small relative to the size of the regressor. Consequently, large changes in estimates of *Net Social Value* are necessary to change the planned number of sales. For instance, the estimated oil spill cost-component of *Net Social Value* is sufficiently small to have essentially no impact on the planned outcome according to the statistical model.

The importance of "in-house" generated information can be evaluated by comparing columns 6 and 3. A joint test fails to reject the hypothesis that only the internal information mattered in the planned decision.

The structure of the full model, column 6, as discussed earlier, includes a mixing coefficient d which is not directly estimated in this model. However, the most significant individual variable in column 6 is a scientific variable— *Net Social Value*. Letting that individual significance select the null hypothesis as d equals 1 (the pure science model) then a joint test comparing column 6 with column 3 rejects the added significance of all additional variables. While the selection of the null hypothesis is somewhat arbitrary in this case,[11] one could conclude from the planning model that the systematic part of the decision was solely an exercise in rational planning.

Recall however, that these results apply to the planned and not to the actual decision. The following sections lead up to reestimation of the models with the actual number of sales as the dependent variable.

Political Changes between Planned and Actual Outcomes

In order to understand more fully the reason for the large difference between planned and actual sales, we examine first a series of important political

---

11. If column 1 is selected to compare the added significance of all variables other than Industry Rank, then a joint test rejects the significance of all added variables concluding that only Industry Rank matters. This conclusion would be at odds, however, with the marginal significance of Net Social Value in the full model.

events: congressional moratoria, litigation, and President Bush's OCS task force.[12]

Original environmental concerns about OCS issues date back to the 1969 Santa Barbara oil spill. Since that time, grassroots opposition to OCS oil and gas development has grown from two main sources: coastal residents (more than 50 percent of the U.S. population resides in coastal counties) and coastal tourists.

Opposition to OCS development exploded early in the Reagan administration, when James Watt was interior secretary. This opposition was due to the Reagan administration's dual strategy of proposing large-scale "area-wide" leasing of OCS lands while simultaneously proposing to eviscerate grants to states under the Coastal Zone Management Act, the Coastal Energy Impact Program, and the National Sea Grant College research program. Reaction to this strategy can still be felt in a widespread public antipathy toward OCS development (Cicin-Sain and Knecht 1987).

The earlier legislative process culminating in the 1978 amendments to OCSLA also took place in a political environment marked by the tension between the concern over oil shortages and the concern for environmental protection. Almost immediately after the amendments were passed, environmentalists and coastal states sought to exclude sensitive areas from consideration in the lease plan process. Initial efforts were directed at further amending OCSLA. In addition, the establishment of marine sanctuaries was seen as a way to preclude oil and gas activity (c.f., Hoagland 1983). The sanctuary approach was successful for small ocean areas. Further amendment of OCSLA was hindered by the overlapping jurisdictions of Congressional committees.

Congressional Moratoria

In 1982, as interest groups mobilized over the plans of Secretary Watt to offer large-scale areas of the OCS for lease, the first congressional moratorium was enacted. Attached as riders to the annual Interior Appropriations Acts from 1982 to 1993, moratoria prevented (and still prevent) the Interior Department from leasing "sensitive areas" specified in the legislation.[13] Initially, each moratorium was intended to provide a one-year hiatus in leasing so that further environmental studies might be conducted. As the years passed, areas subject to moratoria grew, and many areas were rolled over from one year to the next—even after they had been deleted from

---

12. Sources for more details on the political events are: Heintz 1988; Kitsos 1987, 1993; Bolze 1990; Cicin-Sain and Knecht 1987; and Van de Kamp and Saurenman 1990.

13. In 1982 and 1983 leasing was prohibited. Subsequent moratoria prohibited "pre-leasing" activities.

Interior's planning process. Backed by state, local, and environmental constituencies, members from the California and Massachusetts delegations were instrumental in gaining acceptance of the moratoria; most other legislators found little reason to object.

Following an interim period during which Judge William Clark headed the Department of the Interior, Donald Hodel moved from the post of energy secretary to that of interior secretary.[14] In early 1985, Secretary Hodel entered into negotiations with members of Congress from California in order to reach a compromise on an acceptable leasing program.[15] An agreement was reached in July on a plan of limited leasing in some areas and 15-year deferrals in others. However, it soon became clear that some of the areas that had been agreed upon as deferrals were commercially promising.

Secretary Hodel conducted a number of local meetings in California at which he explained reasons for abrogating the agreement (Van de Kamp and Saurenman 1990). This action infuriated the California delegation and local communities in that state, and this mistrust between the Interior Department and coastal states has persisted.[16] Congress attempted to institutionalize the negotiation process between the secretary and members of Congress in the fiscal year 1986 DOI Appropriations Act. A specially selected group of members of Congress met with Secretary Hodel on 16 separate occasions to discuss the imposition of lease stipulations for environmental reasons and proposals for long-term deferrals. The secretary was not required to accept these proposals and, in fact, rejected most of them (*NRDC v. Hodel*, 865 F. 2d 288, [1988] 319).

Although exploration programs had increased during the late 1970s and early 1980s, the oil industry was coming up commercially dry on the frontiers. In the Gulf of Mexico, marginal finds could be supported by existing infrastructure, but in frontier areas (North, Mid-, and South Atlantic, Northern California and Washington-Oregon, and parts of Alaska) finds needed to be much larger to support the construction of infrastructure. In January 1986, the price of oil collapsed, sending the domestic oil industry into a period of retrenchment and making the OCS outside the Gulf, Southern California, and parts of Alaska appear too costly for development. Any possibility of congressional support of expanded leasing or removal of moratoria diminished rapidly with the fall in the price of oil.

---

14. Hodel had been an undersecretary during Watt's tenure.

15. The 1985 fiscal year DOI appropriations bill required Secretary Hodel to enter into negotiations with members of Congress and state and local officials to resolve the long-term leasing status of areas off the coast of California.

16. Bolze (1990) notes that similar negotiations over offshore areas in Alaska did result in the deferral of some areas.

Litigation

The legality of Interior's FYLP process has been contested in the federal courts on three occasions. The first involved Interior Secretary Cecil Andrus's 1980–85 FYLP,[17] although the case was litigated as *California v. Watt* I (668 F. 2d 1290 [1981]), since Watt had become secretary of the interior. In this case, the federal court ruled against Interior, holding that the section 18 requirements for FYLP planning studies were both incomplete and imperfect.

Beginning in 1981, Secretary Watt developed a revised "area-wide" leasing plan, which also was litigated (*California v. Watt* II, 712 F. 2d 584 [1983]). This time the Washington, D.C., Court of Appeals deferred to the administrative actions of Interior and the FYLP analyses. The decisions were upheld. Hodel's 1987–92 FYLP was also litigated (*NRDC v. Hodel,* 856 F. 2d 288 [1988]), and once again Interior won on all counts, except for its analysis of the environmental impacts on migratory species.

Van de Kamp and Saurenman (1990) suggest that it was the outcomes in the 1983 and 1988 decisions which led to litigation by coastal states on individual lease sales and to lobbying for moratoria. Suits were filed by environmental groups and state and local governments, focusing primarily on potential violations of the National Environmental Policy Act's environmental impact statement process (Bolze 1990). Plaintiffs in some of these suits were successful, but in general Interior had an excellent record in federal court, including winning a 1983 Supreme Court decision that focused on the federal consistency requirements of the Coastal Zone Management Act (*Secretary of the Interior v. California,* 464 U.S. 312 [1984]). Thus the imposition of moratoria became one of the most viable strategies for coastal states and environmental groups seeking to achieve their purposes. Subsequently, these groups were successful in obtaining passage of an amendment to the Coastal Zone Management Act that requires a finding by the Interior Department that lease sales are consistent with a state's approved coastal zone management plan—thereby rendering moot the Supreme Court's decision.

Heintz (1988, 231) points out that comments on Secretary Watt's draft FYLP from antileasing interest groups "focused on his failure to give environmental and distributional values high priority." These interest groups continued, albeit increasingly unsuccessfully, to litigate, but by far the most successful strategy was to circumvent the Department of the Interior altogether

---

17. Secretary Andrus began the planning process after the 1978 amendments, but the program was not approved until June 1980. Ironically, Congress requested that the secretary add *more* acreage than the secretary had originally planned to lease. This request occurred during the second OPEC oil supply disruption. The FYLP program was litigated initially as *California v. Andrus.* The name of the case was changed when James Watt was appointed Interior secretary in 1981.

by getting Congressional moratoria incorporated into appropriations bills. Thus except for sales in the Western and Central Gulf of Mexico and in some areas in Southern California and Alaska, the FYLP process was hamstrung by the late 1980s.

Bush Task Force

In 1989, President Bush established a special task force to examine the OCS leasing program.[18] Perhaps no other action demonstrates more clearly the extent to which the OCS leasing process is not simply a rational government planning activity but is part of a broader context of multiparty negotiation. The report of the task force was never published by the White House, but its outcome was a ten-year "Presidential moratorium," issued in January 1990, on leasing off the states of Washington and Oregon, Northern and Central California, Southern California (all but 87 tracts), Southwest Florida, and the Georges Bank.[19] We have been informed by participants in the process that Georges Bank and the Washington-Oregon planning areas were the primary foci of the report and that the development of planning options reflected political input. However, we also have been informed that meetings at the cabinet level with presidential participation focused on substantive environmental risk and oil production information for all areas. Nonetheless, political factors associated with imposing a moratorium rather than promoting leasing can be assumed to be a part of the presidential decision in the political climate following the *Valdez* disaster and the critical conclusion of a National Research Council report on the adequacy of environmental information for decision making in some locations and for some impacts (NRC 1989, 1992).

California is a good example of the political considerations potentially related to a particular OCS decision. Prior to the 1992 presidential election, the Republican Party had been concerned that the results of the 1990 census would lead to a gerrymander (as happened after the 1980 census), forcing the loss of six or seven Republican congressional seats. Republican Party strategists thought that if the gubernatorial victory went to a Democrat, instead of to Republican Pete Wilson, the gerrymander could not be prevented. One result of the presidential moratorium, whatever its cause, was that a very contentious state issue for the Republican candidate had been significantly reduced. A similar situation existed in Florida, although other moratorium areas such

---

18. The 1990 budget proposal (submitted in January 1989) proposed deleting OCS sales off the coasts of California and Florida (Kitsos 1993). Just after the task force was established, the *Exxon Valdez* oil spill occurred. Although unrelated to the OCS oil and gas program, the oil spill sharpened public opposition to offshore oil and gas development.

19. A staff summary prior to the preparation of the final report was printed in the Federal Register (1989).

as the Georges Bank (offshore Massachusetts) could not be associated with any immediate electoral impact.

Kitsos (1993) has observed that although the Republican Party earned short-term political benefits from the presidential moratorium, legislators from some coastal states (especially North Carolina) were concerned about why their state was excluded, while others (Florida, California) were concerned with the possibility that the moratorium might be lifted after the 1992 elections. This led to more extensive congressional moratoria in the 1991 and 1992 Interior appropriation acts. The moratoria in the 1992 act covers all of the Atlantic and Pacific coasts, the Gulf coast of Florida, and Bristol Bay in Alaska.

Analyzing the Actual Outcome

In the broader context of political changes that may constrain or alter the political perceptions of the interior secretary, we consider the determinants of the actual number and location of sales held during this period. This ex post analysis is a reflection of interest group dynamics. The secretary has input into the final outcome only to the extent that the number of sales in planning areas not subject to moratoria can be carried out or canceled, depending on conditions prevailing at the time. We hypothesize that the political influence of the opponents of leasing would in this instance become significant determinants of the actual number of sales.[20] The results of the ex post analysis are reported in table 2 and are discussed below.

The political models are discussed first. Column 1 of the table reports a de minimis political model which includes only *Industry Rank*. Taken alone, that rank continues as a significant determinant of planned sales. Column 2 adds the political variables for groups opposed to leasing. No individual variable is significant, but in important contrast to the analysis of the planned outcome a likelihood ratio test for joint significance demonstrates that, at the 5 percent level, the political opposition variables are significant. We conclude that a purely political model cannot rule out either political side as being determinants of the actual outcome.

However, columns 3 and 4 present the pure competing model of the scientific determinants of the actual number of sales. The results are similar to those for the planned number of sales—*Net Social Value* is a significant determinant of actual sales while the environmental science variable is not significant. We conclude that a scientific planning model considered alone

---

20. A descriptive data analysis indicated high correlations between the existence of a moratorium and our political opposition variables. Consequently we use what seem to be the causal variables in place of a moratorium variable itself.

TABLE 2. Poisson Analysis of Actual Sales

	(1)	(2)	(3)	(4)	(5)	(6)
Constant	2.30	1.74	−1.3	1.5	1.0	1.74
	(0.36)	(4.03)	(0.4)	(1.2)	(1.7)	(4.07)
	0.0	66.5	0.1	22.5	56.8	67.0
Net Social Value			1.0×10^{-4}		3.4×10^{-5}	-0.26×10^{-6}
			(1.6×10^{-5})		(2.6×10^{-5})	(0.32×10^{-4})
			0.0		18.8	99.4
Industry Rank	−0.42				-2.8×10^{-1}	−0.35
	(0.098)				(1.2×10^{-1})	(0.14)
	0.0				2.0	1.4
Environmental Sensitivity		−0.0065		-7.1×10^{-3}	1.0×10^{-3}	−0.0066
		(0.007)		(5.4×10^{-3})	(6.1×10^{-3})	(0.0074)
		36.7		18.7	86.4	37.4
Conservation Members		−0.0069				−0.0069
		(0.01)				(0.011)
		48.6				51.4
League of Conservation Voters		0.80				0.80
		(0.91)				(1.10)
		37.9				46.5
Log likelihood	−15.00	−10.53	−18.58	−37.40	−14.08	−10.53

Notes: Variable definitions are as follows: *Net Social Value* equals net present social value; *Industrial Rank* equals industry preferences, the larger number less preferred; *Environmental Sensitivity* equals relative marine productivity and environmental sensitivity, the higher numbers more sensitive; *Conservation Voters* equals members of the Sierra Club, Greenpeace, and National Wildlife Federation in the state; *League of Conservation Voters* is an index based on a discretized ranking by the League of Conservation Voters for Senatorial votes on Clean Air Act Amendments of 1990.

Standard errors appear in parentheses.

P-values (in percent) under standard errors indicate significance level, for example, 5.0 means significant at 5 percent level.

cannot reject the importance of economic measures in determining the actual outcome, although an environmental science measure is rejected.

Mixed models of political and scientific determinants are reported in columns 5 and 6. In contrast to the results for the planned outcome, the political variable *Industry Rank* is the only significant determinant of the actual number of sales in columns 5 and 6. Also in important contrast to the planned outcome, data generated externally cannot be rejected as being a determinant of the number of sales through a joint test comparing models 5 and 6. As these data (*Conservation Members* and *League of Conservation Voters*) also represent political opposition, we conclude that political forces, both pro and con, exclusively determined the actual outcome. Consequently, in light of the structure of full model with its mixing coefficient *d,* we conclude that a value of zero for *d* cannot be rejected.[21] The data thus indicate that the actual outcome was consistent with a polar decision model solely decided by political influence, noting that both political sides influenced the outcome.

Conclusion

A moratorium on OCS leasing is a blunt instrument of management. The results of our analysis indicate that industry was able to get lease sales in the areas for which it cared most while the political opposition was also able to influence the outcome. *Industry Rank* increased in importance relative to scientific information in moving from planned to actual sales. The political interests of opposition groups also became important in the statistical analysis of the actual outcome compared with the planned outcome. This is consistent with our modeling of the OCS decision process as one of competing interest groups in which the final outcome is not influenced significantly by scientific information. We note here that the analysis of political opposition was made more difficult by the omission of any nonindustry group ranking in the Five Year Leasing Plan.

Our investigations suggest to us that further research could involve closer statistical scrutiny of the congressional politics of moratoria. An area-wide moratorium is the converse of James Watt's area-wide leasing. Just as complete reliance on a market determined number of leases (area-wide leasing) might involve inefficiencies due to incomplete incorporation of social costs in private production decisions,[22] the complete reliance on nonmarket negotia-

---

21. In contrast with the planning model, it is possible to reject a hypothesis that *d* equals 1.

22. Evidence presented by Opaluch and Grigalunas (1984) suggests that private companies adjust their bidding for some social costs related to oil spills. As the private bids must exceed the government's reservation price, it is possible that fewer leases would result from the lower bids.

tion processes might involve inefficiencies due to the rent-seeking behavior of interest groups. Although we do not attempt to calculate its level, we suspect that an inefficient outcome has occurred in the case of OCS leasing.

Some commentators suggest modifications of the OCS leasing process. Kitsos (1993) places blame on the Interior Department for failing to recognize the interests of coastal states in the process. Kitsos suggests that the way to satisfy the concerns of the coastal states is by side payments, also known as revenue sharing. Similar proposals are suggested by Walls (1993), Wilder (1993) and others.

Alternatively, Farrow (1987a) has proposed a market-oriented mechanism to open the auction process to nondevelopment bids after adjusting for development biases inherent in the existing process.

The central issue in the OCS leasing process lies in the parochial interests of the different groups in the distribution of the natural resource and environmental wealth of the OCS. The management of these interests requires developing a process of acquiring information and of making decisions in the context of competing interest groups. The rational planning procedures incorporated in OCSLA may have some value in providing information for all participants. However, they seem to have become primarily a legal requirement that has no more than a marginal effect on the actual outcome of where, and how often, leasing will take place.

REFERENCES

Administrative Conference of the United States (ACUS). 1991. *A Guide to Federal Agency Rulemaking*. 2d ed. Washington, D.C.: Office of the Chairman, ACUS.
Becker, G. 1983. "A Theory of Competition among Pressure Groups for Political Influence." *Quarterly Journal of Economics* 98, no. 3: 372–400.
Bolze, D. A. 1990. "Outer Continental Shelf Oil and Gas Development in the Alaskan Arctic." *Natural Resources Journal* 30 (Winter): 17–64.
Buchanan, J.; R. Tollison; and G. Tullock. 1980. *Toward a Theory of Rent Seeking Society*. College Station: Texas A&M University Press.
Buchanan, J., and G. Tullock. 1962. *The Calculus of Consent*. Ann Arbor: University of Michigan Press.
Cicin-Sain, B., and R. W. Knecht. 1987. "Federalism Under Stress: The Case of Offshore Oil and California." In H. N. Scheiber, ed., *Perspectives on Federalism*. Berkeley, Calif.: Institute of Governmental Studies, University of California, Berkeley.
Dunlap, R. 1989. "Public Opinion and Environmental Policy." In J. Lester, ed., *Environmental Politics and Policy: Theories and Evidence*. Durham, N.C.: Duke University Press.

Coastal States Organization (CSO). 1985. "Coastal States and the U.S. Exclusive Economic Zone." Draft report. Washington.

Cropper, M.; W. Evans; S. Berardi; M. Ducla-Soares; and P. Portney. 1992. "The Determinants of Pesticide Regulation: A Statistical Analysis of EPA Decision Making." *Journal of Political Economy* 100, no. 1: 175–97.

Environmental Protection Agency (EPA). 1990. *Environmental Investments: The Cost of a Clean Environment.* EPA-230-12-90-084. December.

Farrow, S. 1987a. "Lease Delay Rights: Market Valued Permits and Offshore Leasing." *Resources Policy* 13, no. 2 (June): 113–22.

———. 1987b. "Does Areawide Leasing Decrease Bonus Revenues?" *Resources Policy* 13, no. 4 (December): 289–94.

———. 1991. "Does Analysis Matter? Economics and Planning in the Department of the Interior." *Review of Economics and Statistics* 63, no. 1: 172–76.

Farrow, S.; J. Broadus; T. Grigalunas; P. Hoagland III; and J. Opaluch. 1990. *Managing the Outer Continental Shelf Lands.* New York: Taylor and Francis.

Farrow, S., and M. Rose. 1992. "Public Information Externalities: Estimates for Offshore Energy Exploration." *Marine Resource Economics* 7:67–82.

Federal Register. 1989. "The President's Outer Continental Shelf Leasing and Development Task Force." *Federal Register* 54, no. 154: 33, 150–65.

Hamilton, J. T. 1993. "Politics and Social Costs: Estimating the Impact of Collective Action on Hazardous Waste Facilities." *Rand Journal of Economics* 24, no. 1 (Spring): 101–24.

Hazilla, M., and R. Kopp. 1989. "The Social Cost of Environmental Regulation." *Journal of Political Economy* 98, no. 4: 853–74.

Heintz, H. T. 1988. "Advocacy Coalitions and the OCS Leasing Debate." *Policy Sciences* 21:213–38.

Hendricks, K., and R. Porter. 1992. "Determinants of the Timing and Incidence of Exploratory Drilling on Offshore Wildcat Tracts." Northwestern Univ. Dept. of Economics. Mimeographed.

Hoagland, P. 1983. "Federal Ocean Resource Management: Interagency Conflict and the Need for a Balanced Approach to Resource Management." *Virginia J. of Natural Resources Law* 3, no. 1: 1–33.

Holling, D. 1990. *Coastal Alert: Ecosystems, Energy, and Offshore Oil Drilling.* Washington, D.C.: Island Press.

Keeney, R. L., and H. Raiffa. 1991. "Structuring and Analyzing Values for Multiple-Issue Negotiations." In H. P. Young, ed., *Negotiation Analysis,* 131–52. Ann Arbor: University of Michigan Press.

Kitsos, T. R. 1987. "Federal-State Relations in Offshore Oil and Gas Development: Is the OCSLA Working?" In G. J. Mangone, ed., *Coastal States Are Ocean States,* 67–86. Newark: Center for the Study of Marine Policy, University of Delaware.

———. 1993. *Troubled Waters: A Half Dozen Reasons Why the Federal Offshore Oil and Gas Program Is Failing—A Political Analysis,* 291–309. Charlottesville: Center for Oceans Law and Policy, University of Virginia. Mimeographed.

McFadden, D. 1976. "The Revealed Preferences of a Government Bureaucracy: Empirical Evidence." *Bell Journal of Economics* 7, no. 1 (Spring): 55–72.

Magat, W.; A. Krupnick; and W. Harrington. 1986. *Rules in the Making: A Statistical Analysis of Regulatory Agency Behavior.* Washington, D.C.: Resources for the Future.

Mohai, P. 1987. "Public Participation and Natural Resource Decision Making: The Case of RARE II Decisions." *Natural Resources Journal* 17 (Winter): 123–55.

Moody, C. E., Jr., and W. J. Kruvant. 1988. "Joint Bidding, Entry, and the Price of OCS Leases." *Rand Journal of Economics* 19, no. 2: 276–84.

Mueller, D. C. 1989. *Public Choice II.* New York: Cambridge University Press.

National Research Council (NRC). 1989. *The Adequacy of Environmental Information for Outer Continental Shelf Oil and Gas Decisions: Florida and California.* Washington, D.C.: National Academy Press.

———. 1992. *Assessment of the U.S. Outer Continental Shelf Environmental Studies Program.* 3 vols. Washington, D.C.: National Academy Press.

Olson, M. 1965. *The Logic of Collective Action.* Cambridge: Harvard University Press.

———. 1982. *The Rise and Decline of Nations: Economic Growth, Stagflation, and Social Rigidities.* New Haven: Yale University Press.

Opaluch, J., and T. Grigalunas. 1984. "Controlling Stochastic Pollution Events through Liability Rules: Evidence from OCS Leasing." *Rand Journal of Economics* 15 (Spring): 142–51.

Rowley, C.; R. Tollison; and G. Tullock. 1988. *The Political Economy of Rent Seeking.* Boston: Kluwer Academic Publishers.

Stigler, G. 1971. "The Theory of Economic Regulation." *Bell Journal of Economics* 2, no. 1 (Spring): 3–21.

Thomas, L. G. 1988. "Revealed Bureaucratic Preference: Priorities of the Consumer Product Safety Commission." *Rand Journal of Economics* 19, no. 1: 102–13.

U.S. General Accounting Office. 1985. *Early Assessment of Interior's Area-Wide Program for Leasing Offshore Lands.* GAO/RCED-85-66. July.

U.S. Minerals Management Service. 1987. *5-Year Leasing Program: Mid-1987 to Mid-1992—Proposed Final,* April.

Van de Kamp, J. K., and J. A. Saurenman. 1990. "Outer Continental Shelf Oil and Gas Leasing: What Role for the States?" *Harvard Environmental Law Review* 14:73–134.

Walls, M. 1993. "Federalism and Offshore Leasing." *Natural Resources Journal* 33, no. 3: 777–94.

Weingast, B., and M. Moran. 1983. "Bureaucratic Discretion or Congressional Control? Regulatory Policymaking by the Federal Trade Commission." *Journal of Political Economy* 91, no. 5: 765–800.

Wilder, R. J. 1993. "Sea Change from Bush to Clinton: Setting a New Course for Offshore Oil Development and U.S. Energy Policy." *UCLA Journal of Environmental Law and Policy* 11, vol. 31: 131–73.

Part 4
Relations between
Governments

CHAPTER 8

Market versus Government: The Political Economy of NIMBY

Gerald R. Faulhaber and Daniel E. Ingberman

1. Introduction

One of the most vexing of practical social choice problems occurs when a project that benefits society as a whole imposes substantial costs on a single group. For example, we all agree that having our trash collected from homes and factories and disposed of centrally (in a landfill or trash incinerator) is a good idea; the hard part is deciding who will live next door to this potentially noxious facility. Other examples include nuclear waste repositories and half-way houses for recovering drug addicts.

Such a problem is often called a NIMBY: "Yes, this project is a good idea, but Not In My Back Yard." Indeed, modern developed economies appear to proliferate NIMBY problems: technological and economic processes which create choice and value also may create (directly or indirectly) negative externalities, and there is little open space remaining in which to dispose of these bad side effects. Further, technology has in some cases vastly increased the scope of these externalities, as in the case of high-level nuclear waste, the effects of which may last as long as man himself has existed.

A goal of this paper is to compare the outcomes of siting decisions by markets and governments. The institutional details of any application of markets or governments will matter; in this paper, however, we abstract from much detail to focus on the following generic allocation institutions:

1. *Market*. A firm (or firms) negotiates with potential hosts to provide compensation in return for siting a noxious facility. An example of this approach is the siting of solid waste landfills by private firms.[1]

The authors would like to thank Howard Kunreuther and especially Stephen Coate for valuable comments.

1. See, for example, "Trash High, Taxes Low: Bucks Town Divided on Landfill," *Philadelphia Inquirer,* Nov. 18, 1990, p. 1

2. *Government.* A legislative body in which all beneficiaries (and other potential hosts) are represented, chooses a site. In particular, we consider

Decide and Announce.[2] A governmental body chooses the host and announces their decision. This body could be a regulatory agency or a legislature (in which the named host is represented). An example of a NIMBY for which this approach is used is in the siting of the low-level nuclear repository in Pennsylvania.

Legislative Siting with Host Veto. The legislature offers a potential host a compensation package (in dollars or political favors) in order to induce the host to accept the NIMBY. The host is not obligated to take the NIMBY and will only do so if the compensation offered provides sufficient incentive.[3] An example of this approach may be recent legislation regarding high-level nuclear waste repositories.[4]

Siting by legislatures under host veto shares an important feature of market siting: voluntary agreement by the host community. Voluntary agreement implies that a key determinant of outcomes is which agent has *bargaining power.* It is no surprise that the distribution of benefits depends upon whether the host or the legislature/firm has bargaining power; it is a surprise that the size of the net social benefits, i.e., efficiency, also depends upon who has the bargaining power, even when the alternatives are endogenously determined. Further, because different allocation institutions can lead to different outcomes, both in allocative efficiency and the distribution of costs and benefits, the central question addressed by this study is: If society, through its legislature, is choosing among these three institutions that it can commit to using for all future NIMBYs, which allocation institution will be chosen?

This is the constitutional, or metagame, problem. Clearly, if all benefits and costs of all future NIMBYs are known at the time of the metagame, then

---

2. This terminology is related to (but not identical with) that of Kleindorfer and Kunreuther (1992).

3. In some cases, the NIMBY may pose risks to the health of current or future generations of residents. Many object to the notion of monetary compensation to a community that thereby increases its health risk, arguing that only the most desperate would sell off their own and their children's good health. Without coming to a judgment regarding the morality of this transaction, we assume that the noxious facility is to be operated and maintained to assure the health and safety of the surrounding community. The negative externalities generated by the facility are assumed to be non-life-threatening.

4. The Nuclear Waste Repository Act of 1982 (*U.S. Code,* sections 10101–26 [1982]) granted potential host states the right to veto a proposed high-level nuclear waste repository within its borders, subject to override of the veto by both houses of Congress within sixty days. See Jacob 1990 for additional details.

legislators—as community representatives—will favor the allocation institution that yields their community the highest net present value. We believe it more likely that the constitutional stage occurs under a "veil of ignorance," in which community representatives know the likely distribution of benefits and costs but do not know their particular outcomes for all future NIMBYs. Therefore, risk aversion becomes a key factor at the constitutional stage.

The principal results of the paper are as follows.

1. The critical factor in determining the efficiency of the outcome is which agent possesses bargaining power. Whether the site is allocated by a government process or a market process is immaterial to the efficiency of the outcome.
2. The outcome of the metagame (i.e., the allocation institution chosen) depends upon the likely distribution of benefits and costs, with low-variance allocation institutions being preferred ceteris paribus to high-variance ones. This can lead to an allocation institution choice that results in inefficient outcomes.

In Section 2, the model is developed. The major results for the different allocation institutions are derived in Sections 3–4. In Section 5, the metagame is developed and the principal result derived. Section 6 summarizes and concludes the paper and briefly discusses several extensions.

2. The Model

The economy consists of n communities indexed by $i = 1, \ldots, n$; each community[5] behaves as a unitary actor. A NIMBY project is proposed; for concreteness, assume that the project is the collection of the trash of all communities and its disposal at a central landfill. This landfill must be located within one of the communities; whichever community is selected as the landfill site experiences a significant cost in the form of disamenities. After the site has been selected, the government or firm provides the beneficial service (say, trash collection) and uses the noxious facility (say, a landfill) in the host community.

Let z_i = gross benefits from the project by community i, and y_i = cost to community i of being the host, $i = 1, \ldots, n$. This cost y_i includes not only the cost of building, maintaining, and operating the facility, but all dis-

---

5. In practice, of course, communities consist of individual households whose views toward a NIMBY may be quite diverse. However, intracommunity political processes are not the focus of this study.

amenities associated with this facility as well.[6] If the project is not implemented, then the *reversion* project is implemented, in which each community receives net benefits of r_i. The cost y_i of the landfill falls entirely on the host; no other communities are affected by it,[7] and that communities are indexed so that $y_1 > y_2 > \ldots > y_n$.

If the project is not implemented, then each community has the option of implementing the project unilaterally; the costs y_i are sufficiently high that every community prefers the reversion to undertaking the project unilaterally: $z_i - y_i < r_i$. The system is closed in the sense that all costs, including those of host compensation, are borne by the beneficiary communities according to a tax or fee arrangement; no outside subsidies are available. If a project is implemented and compensation C_k to the host k is required, then each community i pays a tax (or fee) of $\tau_i C_k$, if community k is the host, where τ_i is the *tax share* of community i. This tax share is independent of the host[8] and the compensation paid, and

$$\sum_{i=1}^{n} \tau_i = 1.$$

Tax shares are exogenous[9] and based upon contractible information, such as tons of trash removed, community income, or assessed value of community

---

6. In the case of hazardous NIMBYs, communities may have poor information regarding the risks to the community of accepting the NIMBY, and therefore underestimate their own y_i (although this lack of knowledge could as easily result in the overestimation of NIMBY costs). For some practical examples of this phenomenon, see Bailey, "Economics of Trash: Some Big Waste Firms Pay Some Tiny Towns Little for Dump Sites," *Wall Street Journal*, Dec. 3, 1991, p. 1. The role of information in assessing a community's "true" y_i is very important, but not the focus of this study. We assume each community knows its own y_i and acts accordingly.

7. That is, if a landfill is located with a particular community, it is noxious to the residents of that community, but not to that township's neighbors. Indeed, if political boundaries are optimally drawn, this property might define a "community": a land area/population that fully internalizes such externalities. In practice, this need not be true; a community could locate a landfill on its border with another community, thereby minimizing its impact on residents of the host but maximizing its impact on its neighboring community.

8. If this assumption is relaxed, then full optimality can be obtained, both in allocating the NIMBY and in the constitutional choice of institution. If it is possible to design individually tailored payments for each community which depend upon preferences and costs, then net benefits can be equalized across all communities for each NIMBY. This "equal sharing" allocation rule always results in efficient sites and also results in minimum variance among communities. In practice, however, such individually tailored payments are not observed, as they are an obvious invitation to abuse. Fixed tax shares τ_i reduce this potential abuse and are more likely to occur in practice. For this reason, we limit our attention to the fixed tax share case.

9. This assumption reflects our focus on the *level* of compensation rather than the *distribution* of its costs to individual communities.

property. It is sometimes useful to think of the tax share in terms of a "tipping" fee paid for the landfill use. For example, suppose each community i generates G_i tons of garbage. All communities pay a fee to the host community of a per ton, where

$$a \sum_{j=1}^{n} G_j = C_k.$$

Then community i's tax share is simply their fraction of the total garbage:

$$\tau_i = G_i \Big/ \sum_{j=1}^{n} G_j.$$

The project is "all or nothing," so that within each community, either all of the trash is removed or none of it is removed. The net benefit to community $i \neq k$, a non-host, of the project (net of the reversion) is $z_i - \tau_i C_k - r_i$, and the net benefit to community k, the host, of the project (net of the reversion) is $z_k - r_k - y_k + C_k(1 - \tau_k)$. The most *efficient* site is that community k^* which maximizes total net benefits: $k^* = \underset{k}{argmax} \{\Sigma_{i=1}^{n}(z_i - \tau_i C_k - r_i) + C_k - y_k\} = \underset{k}{argmin}\{y_k\} = n$. Since the gross benefits (net of reversion) do not depend upon which community hosts the landfill, the community with the lowest total cost (including that community's valuation of the negative externalities) is the most efficient site.

3. Bargaining Power

A key concept of this paper is *bargaining power*: the ability to be the first to commit to a "final" offer and credibly refuse to bargain further with one or more other agents. This ability to commit is nontrivial: for example, a "final" offer C from a host to a firm may be substantially in excess of that host's costs, presenting potential opportunities for further bargaining. By committing to bargain no more, the host denies itself and the firm these opportunities and confronts the firm with a "take-it-or-leave-it" situation. On the other hand, this form of commitment is limited; agents cannot commit to arbitrary actions. The ability to be the first to make such a commitment is also essential to possess bargaining power. For example, consider a two-player game in which both players have the ability to make "final" offers, but one of the players gets to make his final offer first. Then the power of the other player to make a "final" offer counts for nothing. Bargaining power, then, is possessed by that agent (if any) who is the first to make the last offer; by definition, there can be at most one such agent.

For simplicity, we restrict attention to two polar cases: either the potential host has bargaining power, or the siting institution has bargaining power.[10] We do not model the difficult and important question of how such bargaining power is acquired. The relevant institutions often assign bargaining power: if an agent is able to take a position via a costly action, which can only be undone by another costly action, the agent can thereby achieve bargaining power. For example, communities may be able to obtain bargaining power by adopting land use plans and zoning ordinances which are costly to revise. A legislature may be able to obtain bargaining power by embodying their site choice in legislation which is costly to revise. The focus of this chapter is to demonstrate the importance of bargaining power, not to model how it arises.

The allocative and distributive impact of bargaining power will depend on the information structure. In the case at hand (trash disposal), costs y_i, benefits z_i, and the reversion outcome r_i are all likely to depend upon observable community characteristics, such as land values, income, population density and distribution, even community geology. We thus assume that all information is observable by all parties.[11]

3.1. Host Bargaining Power

Market Processes
Assume that all agents (firm[s], communities) make offers and counteroffers of compensation to take landfills until there remains no mutually beneficial gains from trade. The firm(s) which secure landfills then sell trash collection services to all communities. If there are multiple firms, all have identical technologies; therefore, the firm that has secured the host with the lowest compensation gets all the business. Therefore, at most one landfill operates in equilibrium.

A salient feature of both competitive and monopoly market processes is that the outcome affects not only the "winner," but all other participants in the market, as they are taxed to pay the compensation to the chosen host (or pay for the compensation indirectly through their waste disposal fees). The winning community wants the highest possible compensation and all other com-

---

10. It is easy to show that the case of no one having bargaining power is equivalent to host bargaining power.

11. Information is not necessarily contractible, in that although all parties "know" costs and benefits, it may not be possible to condition court-enforceable contracts on this knowledge.

munities want the lowest possible compensation (and therefore the lowest taxes or waste disposal fees). This important element of the NIMBY market is in sharp contrast to, say, an art auction, in which the losers are indifferent to the price paid by the winner.

PROPOSITION 1. *If information is observable and the potential host community has bargaining power, then either a competitive or monopoly market selects the efficient site n with compensation* $C_n = y_{n-1}$.

PROOF. Community n's best offer is $C_n = y_{n-1}$; no other community can credibly underbid, as all would prefer to face the tax $\tau_i C_n$ than take the NIMBY at a cost $y_i - C'(1 - \tau_i)$, for $C' < y_{n-1}, i = 1, \ldots, n - 1$. The monopolist need not offer more than C_n, and offers of less would not be heeded. Community n's offer is lowest; any firm accepts it and the efficient site is the equilibrium outcome. □

Intuitively, it is clear that neighborhood n prefers to bid y_{n-1}, and the discipline of the market prevents it from bidding higher. It is in the interest of all other neighborhoods to achieve the lowest possible bid from the winner, in order to minimize the resulting tax burden, so that some neighborhood $i \neq n$ may bid $C' < y_{n-1}$ in the hope that neighborhood n would be forced to underbid C'. However, since all information is observable, neighborhood n would realize that such a bid is not credible, in that it would be canceled if a firm accepted it. Therefore, neighborhood n need not underbid such a C'.

Legislative Process
The legislature consists of representatives from every community; decisions are made by majority rule and are binding on all communities. Legislative decisions are made with *sophisticated voting over an endogenous agenda*.[12] That is,

1. All communities take actions based on full knowledge of their consequences.
2. Any representative can make any proposal (consisting of a site i and compensation C_i) at any time; such a proposal is immediately voted against the current proposal, and the winner becomes the new current proposal. This continues until no more proposals are forthcoming.
3. If the current proposal defeats the reversion, it becomes the legislature's offer to the specified community.

---

12. Endogenous agenda means that the order of voting is not prespecified but is determined by the actions of the majority. See Austin-Smith 1987 or McKelvey 1986.

4. If the legislature makes an offer to community k, then k can choose to veto the offer or accept it.

   a. If community k accepts the offer, the project is implemented with the site in k, with compensation C_k and taxes $\tau_i C_k$ for communities $i = 1, 2, \ldots, n$.

   b. If k vetoes the offer, the legislature makes an offer to another community, continuing until either some community has accepted an offer, or no community accepts an offer and there are no more motions forthcoming. In this last case the reversion R is implemented.

An *agenda* $\{(i, C_i), (j, C_j), \ldots, R\}$ is a sequence of communities and compensations: first offer the site to community i with compensation C_i; if i turns it down, next offer the site to community j with compensation C_j; if j turns it down, etc. If all communities turn it down, the reversion is implemented. The *equilibrium agenda* is the agenda that minimizes the total compensation the institution pays.

The equilibrium agenda is assumed to be subject to the following:

A1   The mth agenda alternative must be the institution's most preferred of the remaining alternatives (*subgame perfection*).

A2   The last alternative on all agendas is the reversion R (*"if all else fails . . . "*).

We make the additional assumption that there are at least two host sites for which *every* community prefers at tax shares τ_i to the reversion:

$$z_i - r_i > \tau_i y_{n'}, \text{ for } n' < n \text{ and all } i. \tag{1}$$

Note that this assumption is stronger than the existence of an efficient site; it states that there are at least two sites that, if priced at cost, could command a unanimous agreement versus the reversion.

PROPOSITION 2. *With host bargaining power, legislative siting with host veto results in the efficient site n being selected, with compensation $C_n = y_{n-1}$.*

PROOF. If presented with any offer $C' < y_{n-1}$, community n optimally vetoes the offer, and refuses to accept any compensation except $C \geq y_{n-1}$. It is clear that no community $i \neq n$ prefers to accept the site at $C' \leq y_{n-1}$, given the alternative of that n accepts the site at $C_n = y_{n-1}$. Hence the unique sophisticated voting, endogenous agenda equilibrium is the offer $C_n = y_{n-1}$ to community n, who accepts.   □

Intuitively, legislative siting under host veto shares the two key properties of market siting: (i) voluntary agreement by the host, and (ii) all non-hosts want to minimize the compensation paid to the host in equilibrium. Because there is only one legislature—rather than many legislatures competing for the right to organize trash disposal—the "demand-side" of the "market" with legislative siting under host veto is most analogous to the monopoly market siting described above. Similarly, endogeneity of the agenda (any community can make any proposal) is analogous to the "supply-side" competition between potential host communities in the case of market siting. It is these two properties, shared by both institutions, which lead to identical outcomes for both.

However, this "identical outcomes" result depends critically on inequality (1). If inequality (1) does not hold, siting by market process would always result in the efficient host being chosen, including the reversion if no host is efficient. However, this need not occur with legislative siting. In particular, (*a*) there is a project for which net benefits (over the reversion) are positive in aggregate but not for a majority; or (*b*) there is no project for which net benefits are positive in aggregate but they are positive for a majority. As an example of the latter, suppose a (bare) majority of communities receives most of the benefits from the project, but every community must pay any compensation according to their tax share τ_i. Then a project with negative net social benefit could be approved by this majority, effectively forcing the minority to subsidize the compensation cost. Intuitively, the problem arises because each community has a single vote in the legislature, but can "vote" with dollars in the market.

3.2. Institutional Bargaining Power

We now show that even with a voluntary agreement rule and supply-side competition, an inefficient host can result when the siting institution has bargaining power. For ease of exposition, we treat only the case in which the siting institution is a legislature; identical results obtain if the NIMBY is sited by a private monopolist (again, assuming inequality (1) holds). That is, suppose the legislature can make "take-it-or-leave-it" offers. This *institutional bargaining power* simultaneously permits the siting institution to *structure the alternatives available to a potential host, should it turn down an offer,* and to *make offers which drive any potential host to indifference against the equilibrium alternative.*

The ability to commit to no further bargaining with community i limits the institution to alternatives which are less attractive to both, but particularly to community i. Since it knows what the alternative is, and finds this alternative more unattractive than, say, the next cheapest site, community i is willing to accept less compensation to avoid this alternative.

More formally, the play of the game is as before, with the following additional assumption:

> A3   The institution can commit not to bargain with a community that has refused an offer (*"take-it-or-leave-it"*).

Suppose that the legislature offers community i the site with compensation C_i. Consider community i's decision; it rationally anticipates the remainder of the agenda and can determine the outcome if it refuses its offer of C_i: either some community h will accept with compensation C_h, or the reversion will occur. Therefore, community i will accept its offer iff

$$C_i \geq \frac{y_i - (z_i - r_i)}{1 - t_i}, \text{ if the reversion is the alternative, or}$$

$$C_i \geq \frac{y_i - t_i C_h}{1 - t_i}, \text{ if community } h \text{ with compensation } C_h \text{ is the alternative.} \quad (2)$$

The compensation-minimizing institution offers the minimum C_i consistent with (2). Thus, in contrast to the case with host bargaining power—where the potential host obtains all quasi-rents between its site and the next-cheapest alternative—with institutional bargaining power, the potential host can be driven to indifference against a more costly alternative, thereby resulting in less compensation. In fact, with institutional bargaining power, host compensation is *always* less than its NIMBY cost:

> PROPOSITION 3. *If a legislature with bargaining power is allocating the NIMBY, then the equilibrium host k's net (of taxes) compensation $C_k - \tau_k C_k$ is less than its NIMBY cost y_k.*

> PROOF. Follows from (2); since $\tau_k C_h > 0$ and $z_k - r_k > 0$, $C_k - \tau_k C_k < y_k$. $\square$

Intuitively, when the institution has bargaining power, it can profitably structure the alternatives available to the equilibrium host and effectively strip the host of its quasi-rents. Since these quasi-rents are determined by the utility the host can obtain from the worst alternative the institution can credibly offer, equilibrium site selection is determined by the distribution of z_i's, t_i's, r_i's and y_i's.

Note that the ability to make "take-it-or-leave-it" offers permits the institution to gain by ruling out alternatives. For example, if the institution's agenda offers the least-cost community first, then the next agenda item (by A2) must be community $n - 1$, which may not be the most effective threat to community n. However, the institution can rule this out by making a "final" offer of, say, zero compensation to community $n - 1$ before making an offer to community n. Since community $n - 1$ will not accept this offer, this rules it out of further consideration. If community n is offered second, it can be threatened with offering the NIMBY to community $n - 2$, the best of the remaining alternatives, a more effective threat than community $n - 1$. In fact, all communities can be ruled out in this fashion, so that community n can be threatened with the reversion. The agenda notation does not explicitly include the vacuous offers which are made simply to rule out certain alternatives.

We use the term "*m*-offer agenda" to refer to an agenda in which sites the number of sites that have not been ruled out is m (plus the reversion).

PROPOSITION 4. *If a legislature with bargaining power is allocating the NIMBY, then the equilibrium agenda has no more than two offers.*

PROOF. Let the agenda $A = \{(i, C_i), (j, C_j), R\}$ minimize compensation $C = C_i = \dfrac{y_j - \tau_i C_j}{1 - \tau_i}$ over all two-offer agendas. Therefore, C_j must be the maximal threat against community i. Inserting the subagenda A' after j, $\{(i, C_i), (j, C_j), \{A'\}, R\}$, yields lower compensation only if it increases C_j. However, the agenda $\{(j, C_j), \{A'\}, R\}$ must be subgame perfect, and therefore can only decrease C_j. Since agendas of more than two offers can only increase compensation, they cannot be equilibrium outcomes. □

The intuition behind this result is straightforward: in equilibrium, the institution optimally threatens the host with the largest credible tax bill possible, which comes from offering it to the most costly subgame perfect threat. Adding more alternatives must decrease the cost of this alternative, making the threat less dire, and thus requiring more compensation to the host.

In the case in which all communities realize the same net benefits and face the same tax share, the equilibrium agenda involves only a single offer, with the efficient site chosen.

PROPOSITION 5. *If (i) a legislature with bargaining power is allocating the NIMBY, and (ii) all communities have identical benefits and tax* *shares ($z_i - r_i = z - r$, and $\tau_i = \tau$), then the efficient site n is chosen by a one-offer agenda.*

PROOF. We first show that a one-offer agenda is optimal. Assume the contrary; in order for a two-offer agenda to be optimal for the institution, there must be at least one community j which finds the reversion to be less of a threat than a subgame perfect offer to another community k. Let C_j be the minimum compensation required relative to the reversion, and let C_j' be the compensation required relative to the alternative of community k accepting compensation C_k. Assume the proposition false; then for some j, $C_j > C_j'$.

$$C_j = \frac{y_j - (z - r)}{1 - \tau} > \frac{y_j - \tau C_k}{1 - \tau} = C_j'', \text{ or } \tau C_k > z - r.$$

Thus, for *all* communities i, $\tau C_k > z - r$, and the tax bill for all communities exceeds their benefits net of the reversion. Therefore, offering the NIMBY to k with compensation C_k is not a credible threat against j, as all would prefer the reversion to (k, C_k), and the equilibrium involves a one-offer agenda. The legislature chooses as host

$$\underset{k}{argmin}\ C_k = \underset{k}{argmin}\ \frac{y_k - (z_k - r_k)}{1 - t_k} = \underset{k}{argmin}\ \frac{y_k - (z - r)}{1 - t} = n.$$

Since uniformity across communities ensures efficient site selection in a one-offer agenda, it is clear that with institutional bargaining power, inefficient site selection or multiple-offer agendas can only occur if communities are diverse in the benefits they derive from the project.

The following two examples illustrate how an institution can use its bargaining power and its agenda control to minimize the compensation paid to the host community, and how this power can lead to inefficient site choices.

Example 1

Community	$z_i - r_i$	y_i	τ_i
1	60	130	.333
2	77	120	.333
3	55	100	.333

In this example, the net benefits (192) exceed the cost of any of the sites, yet no community gets sufficient net benefit to take on the project without compensation ($z_i - r_i < y_i$). Consider first simple agendas: $\{(i, C_i), R\}$. The compensation each community requires if the reversion is the alternative is:

$C_1^R = \dfrac{130 - 60}{.07} = 105.00$	$C_2^R = \dfrac{120 - 77}{.67} = 64.50$	$C_3^R = \dfrac{100 - 55}{.67} = 67.50$

In this case, community 2 would be selected host, even though it is not the most efficient. The host is paid a total compensation of 64.50 (compensation net of taxes is $64.50 - \frac{1}{3}[64.50] = 43.00$), which is less than its NIMBY costs of 120.00.

Consider next two-offer agendas: $\{(i, C_i), (j, C_j), R\}$, and let C_i^j = minimum compensation required for community i if community j is the alternative:

$$C_i^j = \frac{y_i - \tau_i C_j^R}{1 - \tau_i}.$$

$C_1^2 = \dfrac{130 - 21.50}{.67} = 162.75$	$C_1^3 = \dfrac{130 - 22.50}{.67} = 161.25$
$C_2^1 = \dfrac{120 - 35.00}{.67} = 127.50$	$C_2^3 = \dfrac{120 - 22.50}{.67} = 146.25$
$C_3^1 = \dfrac{100 - 35.00}{.67} = 97.50$	$C_3^2 = \dfrac{100 - 25}{.67} = 117.75$

In this case, the one-offer agenda leads to the lowest compensation of 64.50 for community 2; all two-offer agendas require more compensation.

However, modifying this example leads to different results; suppose that community 1's cost y_1 is 175 rather than 130.

Example 2

Community	$z_i - r_i$	y_i	τ_i
1	60	175	.333
2	77	120	.333
3	55	100	.333

Compensation required against the reversion differ only for community 1.

$c_1^R = \dfrac{175 - 60}{0.67} = 172.50,$	$c_2^R = \dfrac{120 - 77}{0.67} = 64.50$	$c_3^R = \dfrac{100 - 55}{0.67} = 67.50$

As in example 1, community 2 is the compensation-minimizing choice from one-offer agendas, with required compensation of 64.50. However, two-offer agendas lead to a different result.

$c_1^2 = \dfrac{175 - 22}{0.67} = 230.25,$	$c_1^3 = \dfrac{175 - 23}{.67} = 228.75$
$c_2^1 = \dfrac{120 - 57.50}{.67} = 93.75,$	$c_2^3 = \dfrac{120 - 22.50}{.67} = 146.25$
$c_3^1 = \dfrac{100 - 57.50}{.67} = 63.75$	$c_3^2 = \dfrac{100 - 21.50}{.67} = 117.75$

In this particular case (but not in general), the efficient site, community 3, is chosen. The agenda $\{(3,63.75), (1,172.50), R\}$ is the optimal agenda for the institution. Note that community 3's tax bill, should community 1 accept the NIMBY, is greater than the benefits net of the reversion for community 3: $z_3 - r_3 = 55 < 57.50 = \tau_3 c_1^R$. This is a necessary condition in order that the minimum compensation is realized with an agenda of more than one offer.

The above examples illustrate that if bargaining power rests with a legislature, then (i) the host's NIMBY cost is greater than its net compensation; (ii) inefficient equilibria can result; and (iii) multiple-offer equilibria can result.

4. Decide and Announce

Now consider the case in which the legislature decides upon a site and the host community must take it. There are two subcases of interest: (i) the selected community is forced to accept the NIMBY ("Decide, Announce, Dump"); or (ii) the selected community may be able to pay some other community (in dollars or political favors) to accept the NIMBY ("Decide, Announce, Bargain").

In the case of Decide, Announce, Bargain, assume the legislature assigns the NIMBY to an arbitrary community i. This community's best strategy is to "sell" the NIMBY to the lowest bidder, just as would occur if community i were a private monopolist without bargaining power (see Kunreuther and Portney 1991 for a proposal to implement this procedure using a lottery). In this case, of course, the lowest bidder will be community n. Since the default option is that community i is host, it seems reasonable to assume that community n will have bargaining power in this case. Hence the results of proposition 1 obtain: the efficient site n takes the NIMBY and community i pays compensation $C_n = y_{n-1}$ (which is assumed to be within community i's budget set).

In the case of Decide, Announce, Dump, one might imagine that bargaining also takes place, but as the prelude to the legislative siting decision. If bargaining among communities in the legislature occurs on a cash basis, then we would expect that Decide, Announce, Dump, would produce the most efficient site, just like Decide, Announce, Bargain. Then net benefits to community $i \neq n$ are $z_i - r_i$ and net benefits to community n are $z_n - r_n - y_n$. Although legislative bargaining might be based on a nonmonetary "currency" such as political favors or legislators' votes on other issues, the assumption of no externalities suggests that no community has an incentive to vote against any mutually beneficial trade between community i and community n. In this case, the legislature chooses community n, the most efficient site, as the host.

However, Decide and Announce need not lead to efficient site choices, since the institution does not require voluntary agreement. Since no compensation need be paid for any site, Decide, Announce, Dump lacks any formal mechanism that disciplines the majority not to select arbitrary sites. Nevertheless, the possibility of inefficient site choices under Decide and Announce does not affect the primary conclusions of the paper, as we now show.

5. Constitutional Choice of Allocation Institution

The discussion of the previous sections assumed that the institution for determining the financing and siting of the NIMBY is in place. Of course, the choice of institution generally precedes the need to actually allocate and finance NIMBY projects. The choice of institution is determined in the constitutional stage, or *metagame*, defined as follows:

  I. The legislature (as previously defined) chooses among several options.
     A. legislative allocation
       1. decide and announce
       2. host veto, host bargaining power
       3. host veto, with legislative bargaining power and agenda control
     B. market allocation
       1. competitive
       2. monopoly, host bargaining power
       3. monopoly, with monopoly bargaining power and agenda control.
  II. Communities and their legislators know the *distribution* benefits and costs of the prospective NIMBY(s) $\{z_i, r_i, y_i, \tau_i\}$. Each community

costs of the prospective NIMBY(s) $\{z_i, r_i, y_i, \tau_i\}$. Each community has a uniform prior distribution of its position in this distribution ("veil of ignorance").[13]

III. All communities are risk averse, preferring smaller variance of outcome to larger, *ceteris paribus*.

Several results follow immediately from the definition of the metagame and the propositions of the previous section; we state these results without proof.

RESULT. Since all communities are in identical positions (and everyone knows this), all will vote alike so that decisions are unanimous.

RESULT. Allocation by monopoly market or by the legislature lead to identical results, and depend only upon which agent has bargaining power (Proposition 1 and Examples 1 and 2).

RESULT. Market competition leads to the same outcome as monopoly with host bargaining power (Propositions 1 and 2).

In order to focus more sharply on the role of variance in the metagame, we make the assumption that:

A4   *Net benefits* $(z_i - r_i)$ *and tax rates* (τ_i) *are identical for all communities.*

From proposition 5, this assumption implies that the efficient site n is chosen for all institutions, with the possible exception of Decide and Announce. Of course, examples 1 and 2 show that other distributions of net benefits and tax shares can lead to inefficient outcomes. However, this assumption permits us to focus on the *distribution* of the benefits of NIMBY projects that results from each of the alternative institutions, rather than on the *size* of the total benefits that results. In other words, given assumption (4), the institution chosen in the metagame under the veil of ignorance will minimize the variance of each individual community's payoffs.

In general, if compensation C is paid to the host, then the gamble that each community faces in the metagame is:

---

13. Note that knowing either less or more in the metagame results in trivial outcomes. If *nothing* is known about prospective benefits and costs, then proposition 7 implies that all communities prefer to postpone the metagame pending the receipt of more information. If *everything* is known about prospective benefits and costs, then a decisive coalition adopts Decide and Announce (in either version) and assigns the NIMBY as it chooses.

be host with probability $\frac{1}{n}$ and receive net benefit $z - r - y_n + C\left(\frac{n-1}{n}\right)$;

be a non-host with probability $\left(\frac{n-1}{n}\right)$ and receive net benefit $z - r - \frac{C}{n}$.

The variance of this gamble is

$$\text{Var}(C) = \left(\frac{n-1}{n^2}\right) \cdot (C - y_n)^2 \tag{3}$$

The next proposition compares the frequently used institution of Decide and Announce with that of host veto with legislative bargaining power. Despite the popularity of the Decide and Announce approach in practice, the next proposition demonstrates that Decide and Announce is *never* preferred at the constitutional stage.

PROPOSITION 6. *Institutional bargaining power is preferred by all to Decide and Announce.*

PROOF. With institutional bargaining power (M), the host receives r and non-hosts receive $z - \left(\frac{1}{n}\right)C = z - \frac{r - (z - y_n)}{n-1}$. First, assume that Decide, Announce, Dump (D) leads to the efficient site choice, in which the host receives $z - y_n$ and non-hosts received z. With Decide, Announce, Bargain (B), the eventual host receives $z + (y_{n-1} - y_n)$, the host first selected receives $z - y_{n-1}$, and all others receive z. Therefore,

$$\text{Var}(M) = \left(\frac{1}{n}\right)r^2 + \left(\frac{n-1}{n}\right)\left(z - \frac{r - (z - y_n)}{n-1}\right)^2$$

$$\text{Var}(D) = \left(\frac{1}{n}\right)(z - y_n)^2 + \left(\frac{n-1}{n}\right)z^2$$

$$\text{Var}(B) = \left(\frac{1}{n}\right)(z - y_{n-1})^2 + \left(\frac{1}{n}\right)(z + (y_{n-1} + y_n))^2 + \left(\frac{n-1}{n}\right)z^2$$

Since no community wants to undertake the project unilaterally, $z - y_i < z - y_n < r$, so that

$$\text{Var}(B) > \text{Var}(D) > \text{Var}(M).$$

If inefficient sites can result from Decide, Announce, Dump, then the variance Var(D) is higher and the expected net benefit is lower than hypothesized above.

The intuition suggests that this result is far more general. With Decide and Announce, all costs are borne by a single agent. If "Dump," that agent is community n and the cost borne is $C = y_n$. If "Bargain," then community i bears the cost; if $i \neq n$, that cost is $C = y_{n-1}$, or if $i = n$, the cost is $C = y_n$. With institutional bargaining power (and at least one $\tau_i < 1$), the cost $C' < C$ is shared among more than one community. This constitutes "insurance" against being forced to absorb all the costs, and therefore results in lower variance. All communities prefer the lower variance institutions in the metagame; therefore any institution in which the costs of the NIMBY are shared is preferred by all to either form of Decide and Announce.

Therefore, there are only two relevant alternatives at the metagame stage: (1) competition/monopoly with host bargaining power; and (2) monopoly, with monopoly bargaining power.

The first alternative results in compensation $C' = y_{n-1}$ and the second alternative results in compensation $C'' = \dfrac{y_n - (z - r)}{1 - \tau} < y_n < C'$, assuming that the host community's alternative is the reversion r. Under either alternative, the mean net benefit from the NIMBY project is $z - r - \dfrac{y_n}{n}$.

The next result characterizes the choice between host bargaining power and institutional bargaining power. From this we obtain

PROPOSITION 7. *The outcome of the metagame is competitive/monopoly with host bargaining power iff $C' - y_n < y_n - C''$; otherwise the outcome is monopoly with institutional bargaining power and agenda control.*

PROOF. Since (i) all agents are risk averse; (ii) all have the same diffuse prior over their position in future NIMBYs; and (iii) the efficient site n is chosen with either institution, all communities prefer

$$\underset{C',C''}{argmin} \ \text{Var}(C) = \begin{bmatrix} C' & \text{if } C' - y_n < y_n - C'' \\ C'' & \text{otherwise} \end{bmatrix}$$

which follows directly from equation (3). □

Equation (3) shows that compensation $C = y_n$ is first best (minimum variance) and can be viewed as the benchmark. Each alternative in the metagame deviates from this benchmark; competition or host bargaining power yields excess returns (the quasi-rent $y_{n-1} - y_n$) to the host, so that $C' > y_n$. If one community has a much lower cost of hosting a NIMBY than the next best alternative, that community stands to gain a substantial windfall at the expense of the non-host communities. This results in a large ex ante variance of outcome with this institution. On the other hand, monopoly bargaining power drives the host to indifference against the reversion r_i, and so yields returns which are less than the benchmark $C'' < y_n$. If one community has a very bad reversion r_i, then host veto with monopoly bargaining power can lead to compensation levels substantially less than host costs, to the benefit of the non-host communities. This can also result in a large ex ante variance of outcome with this institution. The effect which is the lesser determines which institution is chosen. In either case, the intuition of Proposition 7 is clear: *the institution with the smaller deviation from the benchmark* C = y$_n$ *has the smaller variance and is preferred by all.*

6. Summary and Conclusions

There exists a wide variety of institutions for allocating NIMBYs among communities: competitive and monopoly markets, and voluntary and involuntary legislative allocations. Of the voluntary institutions (markets and host veto), we have shown that the critical determinant of the efficiency of the outcome is not market vs. government, but which agent has the bargaining power: host community or firm/legislature. If the host community has the bargaining power, then the efficient site is chosen and the host receives quasirents over and above the cost of the site. If the firm/legislature has the bargaining power, then (i) the host receives less than the cost of the site; (ii) the most efficient site need not be chosen; and (iii) the minimum compensation may be achieved by confronting the host with an endogenously determined agenda.

The strong conclusion of the paper that it is bargaining power, not market vs. government, that matters. This conclusion of indifference between market and government must be tempered by noting that this depends critically on inequality (1): all communities prefer at least two sites to the reversion. If this assumption is abandoned, then markets still lead to efficient outcomes, but government need not. Thus, we are no longer indifferent between markets and governments on efficiency grounds, absent inequality (1).

The first sections of the chapter focus on the efficiency of the NIMBY allocation under various institutions; the metagame analysis focuses on the endogenous choice of institution behind a "veil of ignorance," assuming efficient outcomes. Therefore the metagame context addresses distributive rather than efficiency issues. In the case of equal net benefits and tax rates, the institution which minimizes the deviation of returns to the host from the benchmark compensation ($C = y_n$) minimizes variance and thus is the unanimous choice. Perhaps most notable is the result that the involuntary institution of Decide and Announce is dominated by host veto with host bargaining power, despite the apparent popularity of Decide and Announce. The practical difficulties in siting landfills and nuclear waste repositories suggests that the involuntary mechanisms in use today are inappropriate policy vehicles, as indicated by the results of this chapter.

If our finding is that Decide and Announce is always dominated at the constitutional stage by another institution, then why is it so popular as a practical matter? We conjecture that the disparity between fact and theory arises from our "veil of ignorance" assumption. In practice, political decisions regarding allocation institutions hardly take place in complete ignorance, but often (we conjecture) with substantial information about the likely outcome of any siting process. Since the objective of a majority of the legislature is to minimize compensation, this is best accomplished using Decide and Announce *if the outcome of the selection process is predictable at the constitutional stage.* In fact, this appears to occur quite often in practice: the legislature/bureaucracy devises an elaborate (but arguably "fair") procedure to determine where the NIMBY is to be sited, and then plays "Decide, Announce, Dump."

Open issues remain, of course. In practice, disamenity spillovers from a host community to its neighbors appears important. In addition, the simplifying assumption of observable information seems counter to real-world situations in which communities are unlikely to know the costs and benefits of others. Therefore, private information may be the more realistic assumption. However, these potentially interesting extensions are beyond the scope of this paper and must await further research.

REFERENCES

Austen-Smith, David. 1987. "Sophisticated Sincerity: Voting over Endogenous Agendas." *American Political Science Review* 81, no. 4:1323–30.
Jacob, Gerald. 1990. *Site Unseen: The Politics of Siting a Nuclear Waste Repository.* Pittsburgh: University of Pittsburgh Press.
Kleindorfer, Paul R., and Howard K. Kunreuther. 1992. "Siting of Hazardous Facilities," in *Handbook of Operations Research* (forthcoming).
Kunreuther, Howard, and Paul Portney. 1991. "Wheel of Fortune: A Lottery/Auction Mechanism for Siting of Noxious Facilities." *Journal of Energy Engineering* 117, no. 3:125–32.
McKelvey, Richard. 1986. "Covering Dominance, and Institution Free Properties of Social Choice." *American Journal of Political Science* 30:283–314.

CHAPTER 9

Politics and Social Costs: Estimating the Impact of Collective Action on Hazardous Waste Facilities

James T. Hamilton

1. Introduction

Facilities that process hazardous waste generate both externalities and contro-versy. Under current state and federal regulatory regimes, siting a hazardous waste facility involves an extensive process of public examination that few facilities survive. A 1987 nationwide survey found that of 81 recent siting applications for commercial facilities that treat, store, or dispose of hazardous waste (TSDs), 31 had been denied or withdrawn, 36 were still under review, and 14 had received operating permits; 6 of the approved permits had not been implemented, however, because of judicial review or market circumstances. In a separate survey in 1987, state hazardous waste officials reported that 52 percent of proposed commercial facilities were rejected because of public opposition (Mason 1989). Neighborhood opposition to facilities which treat hazardous waste (which the U.S. Environmental Protection Agency [EPA] generally defines as waste that is ignitable, corrosive, reactive, or toxic) springs from a variety of public concerns: threats to property values; odors and nuisances posed by the facility's operation; and health and safety concerns raised by the risk of accidental releases of hazardous waste, such as local groundwater contamination from a dump site (EPA 1979).

Standard interpretations of the Coase theorem (Coase 1960) hold that when property rights are well defined, a firm generating externalities will locate where, *ceteris paribus*, it does the least damage. A growing literature

Reprinted with permission from *RAND Journal of Economics* 24, no. 1 (1993): 101–25. I have benefited greatly from the comments of my dissertation advisers, Richard Caves, Joseph Kalt, and Kenneth Shepsle. Helpful comments were also received from Charles Clotfelter, Sherry Glied, Adam Jaffe, Miriam Jorgensen, John Kain, James Medoff, Nancy Torre Dauphinais, Kip Viscusi, two referees, a coeditor of the RAND Journal, and seminar participants at Chicago, Duke, Harvard, Oberlin, Rochester, Stanford, UCLA, and Washington University. Support from the National Science Foundation and the Duke Law School Olin Program in Law and Economics is gratefully acknowledged.

in economics applies the Coasian framework to the location decisions of noxious facilities (Kunreuther et al. 1987; Kunreuther and Kleindorfer 1986; Goetze 1982; EPA 1982). Sealed bidding processes and other demand revelation techniques are proposed, in which communities bid for compensation to receive facilities such as hazardous waste TSDs that are perceived to pose significant environmental risks. Often, in these auction models the unit of observation is "the community," or individuals who are affected by pollution are assumed to negotiate directly and costlessly with the proposed facility. The benefits of an economic market in location rights are explored without examining the operation of the political marketplace. Becker (1983), however, offers a theory of political competition among interest groups, suggesting that the opposition generated by affected neighborhoods may lead to the same efficient outcome as that of the Coase theorem—the location of the firm in the area where, *ceteris paribus,* it generates the fewest externalities. In this model of politics, the communities that opposed the facility the most would have the most incentive to organize in opposition, so that ultimately the facility would end up in a neighborhood where residents placed less value on the risks posed by its operation.

Is the Coase theorem replicated in politics, so that competition among interest groups in the political system will correct potential market failures, such as the failure of firms to consider externalities in their location decisions? Or are there distinctive elements about collective action which mean that the locations that generate the least political opposition may not be the locations that result in the lowest externalities? A firm attempting to site a facility that generates externalities cares about the amount of *expressed* opposition in the community, which may raise the transaction costs of litigation and regulatory hearings, increase the compensation paid to the community, and increase the facility's operating costs once it is started up. The opposition expressed by a community is a function in part of the value placed on the environmental degradation posed by the facility, which determines the amount of compensation demanded and the willingness to expend resources to mount an opposition. The expressed opposition also depends on the ability of individuals to overcome the free-rider problem and engage in the collective action necessary to oppose the siting through the political process. If areas vary in their potential for collective action and thus their potential to organize politically to express their demands, then a community's *expressed* opposition to a facility may be low, even though residents strongly oppose it. The degree to which a firm internalizes its externalities will thus depend on whether affected communities engage in collective action.

The location decisions of facilities that treat, store, or dispose of hazardous waste offer a natural test of how externalities affect firm location decisions. This article draws upon the capacity expansion plans reported in 1987 to the Environmental Protection Agency by hazardous waste facilities to test whether the potential for collective action in communities affects their probability of being targeted for additional waste-processing capacity. Data from the EPA survey allow one to ask the following: Among those counties that in 1987 had commercial hazardous waste facilities in operation, where were firms planning net expansion of commercial capacity for 1987–92? The probability that a county is targeted for expansion of hazardous waste-processing capacity is modeled as a function of variables relating to the supply and demand for hazardous waste-processing services and variables relating to the compensation for environmental damages posed by a facility's operation. An additional variable is included to measure the likelihood, independent of the value that residents place on the environment, that residents will engage in collective action. The percentage of a county's voting age population that voted in the 1980 presidential election is used as a measure of the potential for its residents to engage in the political activity necessary to translate opposition to a facility into actions that raise a firm's costs of siting.

Section 2 focuses on the importance of incorporating differences in collective action into the Coase and Becker models. Section 3 offers evidence that firms do consider the potential for collective action in choosing among location sites and argues that the literature on political participation suggests the propensity to vote can be used to proxy for potential to engage in collective political action. Section 4 describes the empirical models used to estimate whether a county is chosen as a site for additional hazardous waste capacity. Section 5 presents results showing that the higher a county's potential for collective action, the less likely it is to be targeted for additional hazardous waste capacity and the more likely it will be to have part of its waste- processing capacity reduced. Section 6 concludes that these results demonstrate that in a world of transaction costs and free-rider problems, "politics matters" in the exercise of property rights and the distribution of pollution.

2. Externalities and Collective Action

The Coase theorem provides one argument for why a firm that generates pollution might have to consider the negative effects of its emissions on its neighbors. Consider a regime where property rights are fully defined, transaction costs are zero, and people enjoy the right to be free of pollution. In such a world, a firm choosing its location realizes that it will have to pay its neighbors for the "right" to pollute. In trying to limit its future liabilities for pollution in its location decision, the firm will take into account the physical and demographic characteristics of the surrounding neighborhood that would influence the "cost" of the externalities generated: the number of people affected, incomes, vulnerability of property values, and residents' willingness to pay for environmental amenities. The firm would act as if it were calculating the potential damages associated with each of its locations, for it knows that once it locates and generates pollution it will have to negotiate with and pay compensation to each affected party. Thus, holding other factors constant, the ideal location for the firm would be a liability-minimizing neighborhood characterized by low population densities, incomes, and property values.

The fully defined property rights and zero transaction costs assumptions mean that the costs of a firm's pollution can be easily measured, compensation can be costlessly negotiated with all affected neighbors, and payments can be fully enforced. In this regime, the firm will fully internalize the social costs of its externalities and choose a Pareto-optimal location that takes pollution costs into account. The notion that the invisible hand will lead the firm to its optimal location here is far from surprising, given the strong assumptions in the model. The model itself, however, has come under attack from economists and legal scholars who have raised questions about the impact of wealth effects created by property rights and the potential indeterminacy of bargaining outcomes (Schwab 1989). Perhaps the most frequently voiced objection to the Coase theorem is that in "large-number" cases, transaction costs will render the compensation negotiations envisioned in the exercise of property rights unlikely. As Baumol and Oates (1988) put it, "Even where the number of polluters in a particular neighborhood is small, so long as the number of persons affected significantly by the emissions is substantial, the process of direct negotiation and agreement will generally be unmanageable." Yet large-number cases do give rise to collective action. Local groups may oppose the location or operation of a polluting facility by filing court actions, lobbying local governments, or seeking state or federal intervention. To understand the allocation of property rights in a world with transaction costs, one must thus expand the Coasian story to include collective action.[1]

One hypothesis about the voicing of compensation demands through the process of collective action is that the Coasian solution may simply be repli-

---

1. Objections to the Coase theorem based on the divergence of reality from a world of zero transaction costs show both the aim of the theorem and the degree to which it has been misinterpreted. Coase meant his analysis, which shows that in a world of zero transaction costs, property rights will flow to their most valued use, to be a counterfactual baseline, not a predictive tool. The most enlightening use of his result lies not in debating the validity of his assumptions about a zero-transaction-costs world but in asking why the world we see diverges from it and what this implies. As Coase (1988) himself outlines in a recent collection of his works, "The world of zero transaction costs has often been described as a Coasian world. Nothing could be further from the truth. It is the world of modern economic theory, one which I was hoping to persuade economists to leave. What I did in "The Problem of Social Cost" was simply to bring to light some of its properties. I argued that in such a world the allocation of resources would be independent of the legal position, a result which Stigler dubbed the 'Coase Theorem' " (p. 174). Taking the Coase theorem seriously thus involves asking how the presence of transaction costs may affect the ultimate use of property rights.

cated through the process of political competition. This "Coase in politics" outcome is suggested by Becker (1983). In Becker's model, the competition of political interest groups pursuing their self-interests produces government policies that help correct market failures and minimize deadweight losses. In Becker's model, neighborhoods threatened by a firm generating externalities would form groups to oppose the location of the noxious facility. The influence exerted by these groups in turn depends in part on their political resources, such as their incomes and number of members, and the potential loss or gain involved. The neighborhoods where liability from pollution would be the greatest would thus have the greatest stake in opposing a firm's facility location, while areas with lower liability would exert less political influence in their opposition to the facility. The potential result is a competition of interest groups that ends up with the firm locating in the area where (*ceteris paribus*) its externalities generate the least cost to society. Though Becker's analysis shows it is possible for political competition to lead a firm to choose the site where (*ceteris paribus*) it generates the least externalities, the model does allow for variations in the efficiency of exercising political power that may cause divergences from efficient political bargains.[2] A relevant question thus becomes how important these variations in political action are in determining the outcomes of political competition, e.g., do the prospective costs and benefits of policies largely determine political outcomes, or do potential differences in the exercise of political effort have a substantial impact on political competition. Becker's own conclusions downplay the magnitude of divergences caused by variations in political influence, for he states that his analysis "unifies the view that governments correct market failures with the view that they favor the politically powerful by showing that both are produced by competition among pressure groups for political favors" (Becker 1983). His analysis, which he indicates can be used to explain such policies as pollution taxes, explicitly avoids any specification of the details of the political process. As he explains, "Politicians, political parties, and voters will receive little attention because they are assumed mainly to transmit the pressure of active groups." He observes that some groups may be better at translating their resources into influence and that some groups may be better at policing free riding. Since this model of the policy process takes the group (e.g. pressure group, lobbying association) as the unit of observation rather than the individual (e.g. voter, politician, bureaucrat), potential differences in the likelihood of individuals to engage in collective action are not stressed.

---

2. Using the Becker framework to explore the degree of U.S. government intervention in commodity markets, Gardner (1987) found that factors associated with producers' costs of political organizing and the social costs of redistribution were both important in explaining commodity programs.

Examining the nature of collective action, however, is necessary in order to understand how a firm may consider potential opposition to its siting plan and how its ultimate location may diverge from that in the Coase-Becker world.

Communities facing identical potential losses from the location of a noxious facility may nevertheless differ in the effective opposition they offer to the siting because of differences in rates of political participation.[3] The greater an area's potential for participation in collective action, the higher the firm's expected costs of litigation, lobbying, and compensation, and thus the less likely it will be to locate there. Consider an individual's decision to become involved in the location decision (e.g., through voting, lobbying officials, writing letters to the polluting firm) as dependent on the value of a function $BP - C + D$, where B equals the potential benefit to the individual from the outcome sought, P is the probability that the benefits will result if the individual acts, C is the cost of the action, and D represents consumption benefits for the act (Mueller 1979). The benefits an individual places on stopping the siting of a polluting firm depend on the variables associated with potential compensation in the Coase model. The cost C may involve the opportunity cost of volunteer time or political expenditures and may also vary with the transaction costs involved in political participation. BP is most likely less than C, since the probability P that an individual's action will make any difference to the outcome of the decision is extremely low in large groups of people. Hence political participation may depend not on an "investment" motive but on D, the satisfaction (consumption) an individual derives from participation such as from feelings of community obligation or civic virtue. For a discussion of the individual calculus of voting, see Riker and Ordeshook (1968), Ferejohn and Fiorina (1974), Moe (1980), and Aldrich (1990). Why individuals differ in the degree they engage in collective action remains a central paradox of public choice analysis; participation rates may vary due to differences in transaction costs or in ideological satisfaction from political activity. Acknowledging this potential difference by incorporating it explicitly as an exogenous variable in the location decision of a polluting firm is necessary to

---

3. Olson (1965) identified the fallacy of equating group interests with group action, pointing out that it is individuals calculating their own interests who determine the outcome of collective actions. Individual political acts, however, such as voting in elections or lobbying officials, remain a paradox. Each individual realizes that since her action will hardly be decisive, she can free ride off the political action of others and avoid the costs associated with voting or lobbying. Despite this argument, models of rational choice and empirical investigations of collective action in politics do provide hypotheses about which individuals may be more likely to organize in opposition to a location decision. Examining the demographic variables associated with one form of political participation, turning out to vote, Wolfinger and Rosenstone (1980) found that voting is positively correlated with education, income, and occupation; they also determined that groups which were underrepresented among voters relative to their proportion in the general population include the young, Southerners, and minorities.

understand the net compensation demands expressed through the political process.

Although discussion of public opposition to a facility in a Coasian model is often framed in terms of "compensation demands" by opponents, in reality the costs a firm associates with public opposition are much broader than simply the "compensation" it might have to pay to a community in winning siting acceptance. A firm's anticipation of the price of public opposition from a given area can thus be thought of as an aggregation of the costs imposed on the firm by residents: the costs of participating in extensive regulatory proceedings and court battles; the opportunity costs imposed by the delay in construction; and direct payments to the community in terms of corporate donations and taxes.[4]

Consideration of future liability from the release of chemicals into the area is also part of the anticipated costs of neighborhood opposition. As part of lawsuit settlements and part of attempts to prevent future suits, corporations have begun buyout programs, in which they purchase the homes surrounding hazardous plants and thus create "safety zones" (Schneider 1990). To the degree that collective action is involved in suits arising over the operation of a facility and cleanups after hazardous exposure, the anticipated costs of operating in an area are affected by the degree of political activity even after a facility is operational. Operations in a Superfund cleanup explicitly provide for public participation in the Remedial Investigation/Feasibility Study, which determines the extent and costs of hazardous waste cleanups (Hasan and Simmons 1989). Citizen associations have also been involved in pushing for stricter implementation of regulations at operating TSD sites (Lynn 1987). Amendments to the laws that place *ex ante* restrictions on the operation of hazardous waste facilities (RCRA) and cover *ex post* exposure to liability if operations at a facility go awry (CERCLA) both contain citizen-suit provisions that allow individuals to monitor compliance with waste regulations and bring suit to enforce them. Group action may also raise the potential settlement costs through private negotiations or court decisions if operations at a plant give rise to toxic torts.[5] Thus the potential for collective action can raise the anticipated costs of the operation of a hazardous waste TSD throughout its lifetime.

---

4. The *ex ante* costs of complying with federal permitting and local siting processes are the most obvious costs; in a recent (unsuccessful) effort to win approval for expansion of an existing toxic waste treatment facility in New York, CECOS International estimated that it had spent more than $7 million in designing the site and preparing the case for the permit (MacClennan 1990). For a discussion of state and local laws used by groups opposed to sitings to raise the effective costs of location for hazardous facilities, see Gergits 1987, Andreen 1985, and Bacow and Milkey 1982.

5. Jorgenson and Kimmel (1988) provide a listing of citizen suits brought under RCRA and CERCLA. For a description of toxic torts at hazardous waste sites, see Stever 1988.

Two qualifications should be noted to the role of collective action in opposing facility siting. The actual health dangers posed by the operation of hazardous waste facilities are still a subject of considerable debate (Greenberg and Anderson 1984; Gould 1986). Judging whether public perceptions of these dangers are accurate, however, is not necessary in terms of the following analysis; the existence of public opposition, whether founded on accurate perceptions or not, is sufficient to cause the firms to incorporate political action in their siting criteria. Secondly, the influence of collective action in such decisions must often go unacknowledged in discussions of siting decisions, which are often publicly based on purely scientific criteria. Ironically, when West Virginia explicitly enacted legislation governing solid waste disposal that allowed its Department of Natural Resources to deny waste facility permits if they were "significantly adverse to the public sentiment," the law was struck down by a court which held that such public sentiment could be "unreflective and unreasoned" (Percival et al. 1992).

3. Measuring the Potential for Collective Action

Firms locating hazardous waste facilities will consider the expressed opposition in communities, which depends on both the value that residents place on the environment and the potential for residents to overcome free-rider problems and engage in collective action to express their opposition. This section provides evidence that firms locating noxious facilities do focus on whether opposition to them will be translated into political actions that raise location costs. Based on research on political participation and evidence from siting studies, the potential for residents in a county to engage in collective action is proxied for by the percentage of the voting age population that voted in the 1980 presidential election. I do not use voter turnout here as a measure of residents' willingness to accept or reject a facility nor as an indicator of environmentalism. I use it to represent the likelihood that individuals in the county will engage in collective activity to oppose a facility, a factor that firms consider in estimating whether it is likely or credible that opposition among residents will be translated into political actions that will raise the costs of facility location.

The hypothesis that firms calculate variations in a community's potential for collective action to result in expression of opposition assumes that decisionmakers can predict which areas are likely to produce more political action. Reports prepared during attempts to site noxious facilities offer evidence that the potential for a community to engage in collective action is a factor in the siting of such plants. A report prepared by the private consultants for the California Waste Management Board drew on questionnaire data from municipal officials, a telephone survey of private waste handlers, and a review of the literature on siting opposition to develop a list of political criteria describing communities least likely to oppose a facility. Stressing that while many residents may oppose a facility, not all will translate that opposition into effective political action; the report also noted that "subgroups who may be designated 'least resistant' to the siting of facilities in general may in fact also oppose a particular facility, but it is likely their opposition would be less vocal and less persistent than subgroups in the other category" (Cerrell Associates 1984) Outlining the factors that help identify likely participation in active opposition in siting battles, the report concluded, "Candidate sites can be suggested partly on the basis of neighborhoods least likely to express opposition—older, conservative, and lower socioeconomic neighborhoods."

Additional evidence of the focus on the political power of facility opponents comes from a report prepared for a body attempting to site a low-level radioactive waste repository in North Carolina (Epley Associates 1989). Community assessments in the study of likely sites included detailed discussions of current levels of political activity in each area, the strength of neighborhood organizations and local political representatives, and information on the population such as percentage of registered voters by party affiliation. A report prepared for the EPA on public opposition to siting hazardous waste facilities in particular concluded that in examining a community, "One pre-existing condition of particular importance is the political sophistication of the population in the vicinity of the proposed or operating site. If the area has a history of political activism . . . or previous experience opposing a facility siting attempt . . . , then it is likely that organized opposition to the siting attempt will appear that much sooner" (EPA 1979). The largest company in the hazardous waste industry has gone beyond relying on publicly available demographic and historical indicators to predict opposition; it now conducts surveys in communities before it files plans for a new facility, in order to gauge likely opposition (Pearl 1991).

Though evidence from battles over siting controversial facilities such as hazardous waste TSDs shows that firms attempt to calculate differences in community potential for collective action, the question arises of how to proxy the firm's estimate of a community's potential for political opposition. I have chosen a single variable, the percentage of the voting-age population that voted in the 1980 presidential election, to measure the potential for residents to overcome free-rider problems and engage in collective action. Part of the justification for this variable selection is based on case-study evidence. In the political assessment of North Carolina counties targeted as sites in 1989 for a low-level radioactive waste repository (Epley Associates 1989), the consultants hired to conduct the study included for each target county the percentage of voting-age population that voted in the 1980 presidential election. This figure was contrasted with the average turnout in the state, thus indicating whether the population was more or less likely to engage in political action.

A stronger argument for the use of voter turnout figures to proxy collective action is that research in political science indicates that those individuals more likely to become involved in community political action in general and environmental activism in particular also report a higher tendency to vote than others. A study of the populace in the Three Mile Island (TMI) community found that activists in the area (e.g., those who opposed through political action the nuclear plant's continued operation) reported a higher voting percentage compared to those who were free riders (e.g., opposed the plant but did not take part in political action) (Walsh and Warland 1983). People who ended up taking part in actions to oppose the nuclear plant's operation were more likely to vote in the presidential election both before the accident (1976) and after it (1980); this indicates that voting figures in a presidential election can be used as a proxy for likely political action when a local environmental conflict arises. In a logit analysis predicting whether an individual participated in antinuclear activities in the TMI area, Walsh and Warland found that in a model including such variables as education, assessment of the dangers posed by nuclear power, and an individual's political ideology, general political participation (measured by voting in the 1976 presidential election) was a significant predictor of whether an opponent of the facility actually participated in actions to encourage shutdown of the plant after the accident. These results are consistent with the finding that individuals who on average are active in local community affairs or participate in political campaigns report higher average voting percentages (Verba and Nie 1972).

Though reports prepared by political consultants in siting battles and the academic literature on individual political participation provide support for using voter turnout to proxy for collective action potential, this is the first academic work on facility siting to use voter turnout as a measure of a community's potential to engage in political action. Concerns about using the voter turnout variable include the accuracy of such a measure and whether there are better indicators of potential political action. One alternative measure would be the percentage of the population that belongs to environmental groups. But this measure was rejected since it combines two factors, the value placed on the environment by individuals and the likelihood that individuals will not free ride and will choose to participate in politics. Since I hope to isolate the latter factor in my empirical tests, I have chosen voter turnout in a presidential election as a measure of general political activity rather than a more specific measure of political activity that is linked to preferences on a given issue. Those people who join national environmental groups also report voting in higher percentages (Fowler and Shaiko 1987), which raises the question of whether voter turnout is correlated with demographic characteristics associated with environmentalism. At the county level, however, the variables that would be important in predicting the amount of compensation a community might demand for environmental damage (which include housing prices, income, and education) are not highly correlated with voter turnout. Of those variables in the data set that do have statistically significant correlations with voter turnout, the magnitude of the correlation is not large: voter turnout and percentage of the county with a university education have a correlation of .25, while the county's nonwhite population and voter turnout are the most highly correlated at $-.33$. Voter turnout in the presidential election in 1980 has a correlation of .82 with turnout in the 1972 election, suggesting that the choice of 1980 as the particular presidential election to proxy variations in potential political activism results in a measure correlated with other years' indicators of political activity.

The ability of a community to halt the siting of a hazardous waste facility may depend on the actions of both community residents and elected officials. The relative political power of local representatives in city government or state legislative bodies may influence where noxious firms are likely to locate. In states where siting battles are fought with intervention from state-level officials, those districts represented by legislators who are committee chairs or who have more influence with the executive branch may be less likely to be chosen as sites. The influence of elected officials independent of the actions of their constituents is only indirectly accounted for here, however, because of the difficulty of measuring these variations. Elected officials are generally assumed here to be agents who respond to the preferences of their constituents (who as multiple principals may have conflicting stands on a proposed facility siting). Voter turnout may also indicate the potential for opposition to be expressed by politicians, however, if one views it as a measure of how closely the behavior of officials will be monitored. Politicians may be likely to enjoy some slack in the principal-agent relationship if electoral constraints are lower, which may be the case if they are retiring or are not facing another election for several years (Kalt and Zupan 1984, 1990). One might assume that in areas with lower voter turnout, elected officials might be freer to follow their own ends instead of those of constituents who opposed the facility. In some cases, compensation figures paid by the firms may even end up with the officials rather than their constituents (Cray 1991). In this case low voter turnout in an area could indicate that officials there might be more willing to accept a facility because their actions are less likely to be monitored and constrained by constituents who oppose the facility.

4. Data and Methodology

Prior to the 1970s, firms that generated hazardous waste could choose their locations without the constraints of state siting restrictions or federal standards. In 1976, Congress passed the Resource Conservation and Recovery Act (RCRA), which established permitting, record-keeping, and operation standards for facilities that generate, transmit, store, or dispose of hazardous waste. In 1980, Congress passed the Comprehensive Environmental Response, Compensation, and Liability Act (CERCLA), establishing the Superfund program to aid in the cleanup of uncontrolled waste sites. Nearly all states have adapted their own hazardous waste programs, many of which contain provisions that attempt to ease the not-in-my-backyard (NIMBY) problem in siting new disposal capacity. The information generated by these regulatory programs offers the chance to examine both how externality-generating firms located in an era of less stringent regulation and how the advent of stricter controls has affected the distribution of these facilities.

The facilities that treat, store, or dispose of RCRA hazardous waste can be divided into a number of categories: private firms that generate and manage waste on site; commercial facilities that treat or dispose of waste from others for a fee; limited commercial facilities that accept offsite waste from only a few sources (often from a facility under related ownership); public disposal centers (e.g., municipal facilities); and public generators who manage waste onsite (e.g., federal defense facilities). In the data described below collected by the EPA, in 1986 there was a total of 251 commercial TSDs, 258 limited commercial facilities, and 1,071 major generators that processed their waste on site. This article concentrates primarily on a subset of the TSDs, the private commercial facilities whose primary business is treatment or disposal of hazardous waste shipped from other generators.[6] These facilities generate the most public opposition, in part because their externalities are concentrated in the surrounding area without the offsetting benefits of substantial jobs and tax revenues that onsite generators that are manufacturing plants may bring to a community (EPA 1979).

Research relevant to studying the location decisions of hazardous waste TSDs falls into three main areas: analysis of siting procedures for locally undesirable land uses; research on the spatial distribution of pollution among neighborhoods; and examinations of the impact of environmental regulatory costs on firm location decisions. The stalemates generated by attempts to site facilities with concentrated costs and diffuse benefits have generated a large literature on improving the path of public and private negotiations associated with facility sitings (Duerksen 1983; Lake 1987; Morell and Magorian 1982; O'Hare, Bacow and Sanderson, 1983). Sullivan (1987, 1990) and White and Wittman (1981) provide formal models of the distribution and operation of such polluting facilities. Studies of facilities that treat hazardous waste

---

6. There are approximately 650,000 generators of hazardous waste, 2 percent of which account for over 99 percent of the total waste produced (McCarthy 1987). The EPA estimates that 275 million metric tons of hazardous waste regulated under RCRA were generated in 1985, of which 96 percent was managed on site at the source of generation (EPA 1987).

provide numerous case studies of local opposition (EPA 1979) and detailed descriptions of the operation of these facilities (Goldman, Hulme, and Johnson 1986; Office of Technology Assessment 1983, 1986). Focusing specifically on the role of political opposition in siting, Mitchell and Carson (1986) propose that communities be given the right to refuse noxious sitings and that referenda be used to indicate community acceptance of proposed compensation packages for sitings. Regulatory research has focused extensively on both the implementation of stricter controls on hazardous waste handling (Fortuna 1986; Davis and Lester 1988) and the progress toward cleaning sites contaminated by past exposure to hazardous waste (Acton 1989; Office of Technology Assessment 1985).

Though no empirical studies to date have modeled the location decisions of TSDs, three studies do provide a description of the neighborhoods surrounding firms generating or disposing of hazardous waste (Gould 1986; Commission on Racial Justice 1987; Greenberg and Anderson 1984). Gould found that communities with the highest income had the least amount of toxic waste generated. The commission report found that when communities with commercial hazardous waste facilities were compared with their surrounding county, the community with the facility had a higher minority percentage, lower household income, more sites contaminated by previous exposure to hazardous waste, lower house values, and higher levels of waste generated per person. Greenberg and Anderson found that communities with abandoned hazardous waste sites were more likely to have a higher percentage of the poor, elderly, and minorities than other communities. Note, however, that these studies were intended more to describe the exposure of different groups to potential harm rather than to model firm location decisions. They make no attempt to determine when the facilities examined actually located in a specific area, what the characteristics of the neighborhoods were at the time of location, and the influence that the facilities might have had on the area (e.g., facility locates, environment deteriorates, and low-income residents are attracted to the area by falling prices caused by diminished environmental amenities).[7] The role that environmental externalities actually play in firm

---

7. For more direct studies of the influence of potential environmental costs on firm location, see Duerksen 1983, McConnell and Schwab 1990, and Pashigian 1985. Examining data on job growth from *County Business Patterns* reports during the 1970s, Duerksen found in comparing an index of environmental regulation with shifts in industrial employment that states with "lax" environmental laws were not more likely to attract pollution-intensive industries. After analyzing the locations of 50 new branch plants in the motor vehicles industry chosen 1973–82, McConnell and Schwab found only mixed evidence that regional differences in environmental regulation had an impact on factory siting. Pashigian, however, finds indirect evidence on firm location in his study of congressional votes in which members of Congress act as if environmental regulations will influence the pattern of industrial location; he posits that locational competition

location decisions thus remains an open empirical question, one which is modeled in this section in the context of hazardous waste facility plans.

Under 1986 Superfund legislation, Congress required that each state submit to the EPA a Capacity Assurance Plan (CAP) detailing its plans to ensure adequate capacity to deal with hazardous waste generated in the state over the next 20 years. In 1987 the EPA conducted a national survey of hazardous waste treatment, storage, disposal, and recycling facilities (the TSDR survey) to develop a database to support the states in the construction of their capacity reports. The TSDR survey conducted by the EPA in 1987 was a census of all facilities with final or interim status permits issued under the RCRA program to treat, dispose, or recycle hazardous waste. The survey was also sent to a sample of facilities that store hazardous waste, though storage capacity is not contained in the data set analyzed. Firms were required to furnish data on total 1986 capacity to handle waste and amount of this capacity that was utilized, with capacity broken down by different treatment technologies. Respondents were also asked planned maximum capacity for 1987, 1988, 1989–90, and 1991–92. The net future additions to capacity reported for each facility were calculated from these plans. The planned capacity figures have been used by states in calculating the plans they must submit to EPA which certify that they have adequate capacity to deal with their hazardous waste. The projected capacity figures submitted by the firms have also been used by EPA in the promulgation of hazardous waste regulations (EPA 1988). Describing the collection of these figures, the agency noted that "facilities were asked to report any treatment processes planned to be operational (considering construction and permit time) by January 1992" (EPA 1988). Thus, firms were asked to take into account the permitting process, where collective action may slow or halt a siting, in their projections of operational capacity.

The planned capacity changes reported by these facilities represent a unique opportunity to test theories of externalities. Previous attempts to examine the relationship between active facilities and their current surroundings have failed to control for when the facility was sited and how the area may have changed since then. The planned capacity changes in the TSDR survey are all for the same time period (1987–92) and were estimated at the same time (1987), so data can be assembled on the characteristics of the areas considered around the time the plans were made. Furthermore, the fact that data were available at the time only through the use of the Freedom of Information Act means that "announcement effects" of changes in neighbor-

among different regions of the country in part explains support in Congress for regulations that restrict economic activity in areas of the country whose air quality already is better than minimum environmental standards.

hood variables based on reactions to the plans are probably minimal, since the plans had not necessarily been announced in the affected areas.

The survey is not a census of all planned changes in hazardous waste capacity, for to be included in the TSDR panel a facility had to have an RCRA permit or an interim RCRA permit for processing hazardous waste. The database thus misses some planned facilities at new sites that have not entered the RCRA permitting process. This means the database does not represent the location plans of all new hazardous waste facilities; the survey does capture the expansion of existing "facilities," which in reality may involve construction at a new site. Note, however, that expansion of facilities often involves the same permitting and hearing processes as location at a pristine site, so that the same calculus outlined of the potential costs of collective action applies in the firms' determination of where to locate new capacity. Collective action may thus be a potential factor in the decision of which capacity expansions may generate the least opposition and which capacity closures may reduce costly public battles with area residents. To the degree that capacity changes are thought to generate less opposition than the siting of new facilities, the following tests will be biased against finding any evidence of the influence of collective action in the siting process.

A commercial hazardous waste facility i considering whether to expand at location j will maximize the profit function $\Pi_{ij} = f(X_j, N_i)$, where X_j is a vector of characteristics associated with location j and N_i is a vector of firm-specific characteristics. Assuming that firms are homogeneous simplifies the specification of the profit function by eliminating N_i. There remains the problem, however, that the estimated profits a firm making its capacity decision associates with each potential location are unobservable. What is observable is whether a location is chosen as the site of a commercial hazardous waste facility. Assuming that these facilities choose to expand in geographic markets where they can earn at least normal profits, one can estimate the probability that expansion will occur in a given location as a function of the underlying variables X_j that determine the (unobservable) potential profits in the area. For each location j, the probability that a TSD is planning to expand there can be modeled as $P_j = F(\alpha + \beta X_j)$, where F is the cumulative logistic probability function which can be estimated with a logit specification.[8] The location-specific characteristics that determine the potential profitability of capacity changes in a given county, X_j, include the current levels of processing capacity and waste generation, factor costs, potential compensation, and the ability of residents to use collective action to translate compensation demands and opposition into costs the firm must consider. Each of these sets of variables is

---

8. See Carlton 1983 for the application of McFadden's (1973) conditional logit model to the empirical estimation of firm location decisions.

TABLE 1. Testing the Equality of Means of Counties with and without Planned Commercial Hazardous Waste Capacity Expansion

Variable Means		Counties without Expansion (N = 84)	Counties with Expansion (N = 72)	t-Stat.
Hazardous waste processing capacity surplus (M tons)	CAP	2.41 (8.3)	1.03 (1.7)	1.48
Hazardous waste generation (M tons)	WASTE	.39 (1.3)	.46 (1.2)	−.32
Manufacturing value added (M$)	MANUFAC	1,270 (1,750)	1,880 (3,970)	−1.17
Value of land and buildings per farm acre ($)	LAND	2,570 (6,930)	2,910 (9,370)	−.26
Median house value ($)	HOUSE	43,910 (15,510)	48,160 (21,460)	−1.40
Median household income ($)	INCOME	17,320 (3,650)	17,270 (3,310)	.09
Percentage of adults with 4 or more years of college	%UNIV	.15 (.06)	.15 (.05)	−.51
Nonwhite population percentage	%NWHITE	.15 (.15)	.16 (.13)	−.70
Urban population percentage	%URBAN	.69 (.28)	.70 (.27)	−.30
Population per square mile	DENSITY	830 (1,640)	890 (1,930)	−.22
Percentage voter turnout in presidential election	%VOTE	.56 (.076)	.52 (.076)	3.4*

Note: Sample consists of all 156 counties that had commercial hazardous waste facilities operating in 1986. Standard deviations are in parentheses.

*Statistically significant at the 1 percent level.

described briefly below (see table 1 for summary statistics). Unless noted otherwise, county-level demographic data are 1980 figures from the *County and City Data Book, 1983;* all hazardous waste variables have been constructed by aggregating facility-level information obtained from the EPA.[9]

---

9. Data on hazardous waste processing capacity came from the EPA's 1987 *National Survey of Hazardous Waste Treatment, Storage, Disposal, and Recycling Facilities,* obtained by the author under the Freedom of Information Act (FOIA). Figures on waste generation are from the EPA's *1985 Biennial Report Data,* obtained under the FOIA. Age of facility information came from the EPA's *Hazardous Waste Management System Notification Extract,* obtained under the FOIA. Additional information on generation of toxic waste incorporated figures from the EPA's Office of Toxic Substances, *Toxic Release Inventory, 1987.* All census data were obtained from the U.S. Department of Commerce, Bureau of the Census, *County and City Data Book,* 1977 and 1983 editions.

Waste Generation and Capacity

The expansion decisions of commercial TSDs within a county depend in part on current capacity utilization, potential growth in demand from waste generators in the area, and the capacity available in surrounding counties. The size of the current capacity surplus (*CAP*) in the county is measured by subtracting the maximum available annual hazardous waste processing capacity reported for 1986 from the actual amount of waste processed at facilities in 1986; these figures are from the EPA's 1987 TSDR survey. The expansion decisions of private and commercial onsite processors are examined here without the inclusion of the plans of government facilities, since their freedom from exposure to certain types of liability and general operation under a different set of incentives mean they might operate under a different site selection process. County capacity figures do include figures from both private and government facilities; waste generation is also calculated by adding up totals from private and public generators. Capacity figures for landfill technology were estimated in the EPA survey on a lifetime basis rather than an annual basis, which made them less comparable to capacity estimates for other waste-control technologies which do not "fill up" over time as landfills do. Because of this, two separate samples were constructed, one with landfill technology included in the sample and one without. The results reported are for capacity figures without the landfill technology, though all analysis was also done with landfill technology figures and the results (except for figures directly relating to estimates of capacity) were similar. The exclusion of landfill technology causes the number of counties with commercial hazardous waste facilities to drop from 159 to 156, because of facilities which only had landfill capacity. The capacity surplus variable thus represents the difference between the maximum amount of waste that can be processed at commercials TSDs in a county in a given year under existing permits and the amount of waste actually managed at these facilities during the year. Assuming that projections of future demand are based in part on current demand, one is left with the decision on how to measure generation of hazardous waste by potential customers for a county's commercial TSDs. The results reported here use the measure *WASTE,* the total annual amount of hazardous waste generated by facilities in a county as reported in the EPA's 1985 Biennial Report database (which covered approximately 49,000 generators and 5,200 TSDs nationwide).[10]

---

10. Alternative measures of waste employed included data on toxic emissions from facilities in EPA's 1987 *Toxic Release Inventory.* A broader definition of local waste generation, the amount of waste generated in counties within a radius of 280 miles, was also employed. Results in each case did not vary from those reported. The amount of waste generated in the county may also proxy for the potential political support by generators for attempts to site a hazardous waste TSD. To the extent that firms that generate waste become involved in the siting process, they may

Factor Prices

A primary factor whose costs varies with location for a commercial hazardous waste facility is land. *LAND*, defined as the average value of land and buildings per acre, is used to proxy this factor cost. *MANUFAC*, the value added in manufacturing in the county, is also included, in part to capture other characteristics of factor market prices that favor industrial location in the area. *MANUFAC* also serves as a measure of industrialization, which may capture the potential for waste generation that may not appear in the EPA databases.

Potential Compensation

A firm in a Coasian world trying to minimize the compensation arising from capacity expansion would consider how areas differ both in the value they place on the environment (which relates to the amount the areas may be willing to expend in fighting the facility in the *ex ante* regulatory process) and in the amount the firm would have to pay under *ex post* liability laws such as Superfund and those relating to toxic torts if operation of the facility results in chemical releases. A number of methods have been used to test for differences in the value people place on environmental amenities: survey data of individual voters about the location of a noxious facility (Fischel 1979); election returns in an environmental referendum (Deacon and Shapiro 1975); survey questions about willingness to pay for environmental cleanups or "environmentally sound" products (Roper Organization 1990); surveys on the demand for distance from a hazardous waste landfill (Smith and Desvousges 1986a, 1986b); and evidence from hedonic housing-price models of people's willingness to pay for clean air (Harrison and Rubinfeld 1978). Overall the results suggest that residents with higher incomes or higher education levels are willing to pay more for a given change in the level of environmental amenities. Hence demographic variables such as income and education are included as variables relating to Coasian compensation demands.

In addition, the firm may be thought of as calculating the potential costs from court actions arising from release of hazardous chemicals. Variables that would go into court calculations of net present values of discounted earnings streams in damage claims are thus included, such as income, education, race, and poverty status. Value of the physical property put at risk by the potential contamination is captured by the median house value (i.e., as the value of the

---

help overcome opposition to the location of a facility that treats the waste they produce. I am indebted to an anonymous referee for this point.

housing stock rises, the cost of buying out neighbors whose property is damaged by potential contamination increases). The variable *%URBAN,* the percentage of population which lives in urban areas, is included since counties with more rural areas might have more dispersed populations and hence have areas where fewer people might be affected by the location of the commercial TSDs. *DENSITY* similarly captures the number of people put at risk by a potential chemical release.

Collective Action

The greater the potential for collective action in an area, the higher the expected costs of litigation, lobbying, and compensation a firm will face and thus the less likely it will choose to expand in a given area. The potential for residents in a given area to participate in collective action to oppose a firm's location is represented in the models by their voter turnout rate in the 1980 presidential election. The percentage of the population 18 and above who voted (*%VOTE*) is included not because one envisions that a firm concerned with political opposition would necessarily examine such figures; the variable is meant rather to be correlated with the potential for political action that a firm may estimate using a number of factors in its consideration of areas for expansion.

Alternative Explanations

Before examining the results of the empirical models of capacity expansion, questions need to be addressed relating to land prices and environmental quality. A firm may choose not to locate in a neighborhood because land prices there are higher relative to other locations. Since land prices may capitalize the value of public goods associated with a given community, however, the land price may be "high" because political activity by residents resulted in strict zoning laws that protect the environment. The increased security of the community's environment thus results in a higher land price, reflecting the greater "amenity" value of the community's secure environment. If land prices were high in part because of the security provided by effective political action, then multicollinearity could be a problem; at the county level, however, *LAND* and *%VOTE* have a statistically insignificant correlation of only $-.07$.

Note that housing prices may also reflect the disamenity of a firm's location if a hazardous waste facility does locate in a given area. This makes the following story plausible: a firm locates in a given area; externalities from the firm harm the environment and cause the prices of surrounding houses to drop; the low housing prices attract low-income residents who are willing to trade off environmental quality (a normal good) for low housing prices; and the result *ex post* looks as if the firm located in a "low income" area (see Harrison and Stock 1984 for a discussion of the effects of hazardous waste on housing prices). Unless one believes that the location of a firm results in a tipping phenomenon that attracts many more noxious firms or that the stigma of a firm's location affects the entire community, using figures for the county where a firm is located rather than just the characteristics of the blocks surrounding the plant should help lessen this problem of *ex post* changes in neighborhood housing values. Using the plans (some of which may be unannounced) of capacity expansion rather than actual expansions helps avoid the problem of location effects on property prices.

Another plausible explanation for the location of TSDs in low income areas is that if damage is lower in terms of potential liability (e.g., present value of lost income or property values) or the value placed on amenities (measured by residents' willingness to pay for amenities as reflected through housing purchase prices), then a firm offering compensation to prospective neighbors will receive the lowest compensation demand from low income areas (in a Tiebout-style process in which people with low values on environmental public goods wind up situated in the same locality). This compensation may be in the form of tax revenues that reduce residential taxes, anticipated job creation and the multiplier effects in the area, or specific payment for local improvements (e.g., additions to firefighting equipment). The compensation story is often part of the strong community acceptance of onsite processing of hazardous waste at large manufacturing plants that generate employment and tax revenue (EPA 1979). Commercial TSDs, however, generate few jobs in a county; moreover, until recent state siting legislation, TSDs were often free to locate without explicit compensation. If this "low compensation bid" scenario is the reason that hazardous waste facilities choose to expand in a low-income area, however, then again the political variable should have no additional explanatory power.

5. Results

Commercial Facility Capacity Plans

The TSDR survey allows one to examine the impact of collective action on the decisions of hazardous waste firms by exploring the question of which counties with commercial hazardous waste facilities in operation were targeted by firms for a net expansion of commercial capacity for 1987–92.[11] Dividing the

11. When counties with commercial hazardous waste facilities ($N = 156$) are compared to the rest of the counties in the United States ($N = 2,951$), counties with commercial facilities appear to be those with active, prosperous manufacturing centers. They tended to have more

TABLE 2. Determinants of the Planned Expansion of Commercial Hazardous Waste Processing Facilities, 1987–92

Variable	(1)	(2)	(3)	(4)
Hazardous waste processing	−1.3e−7**	−1.5e−7*		
capacity surplus (tons)	(7.0e−8)	(7.0e−8)		
Hazardous waste generation			−6.6e−8	−8.4e−8
(tons)			(1.5e−7)	(1.6e−7)
Manufacturing value added	.0002	.00038**	.00006	.00016
(M$)	(.00014)	(.000218)	(.00008)	(.000127)
Median house value ($)	.00001	.00001	.00001	.00002
	(.000016)	(.000029)	(.000015)	(.000027)
Percentage voter turnout in	−10.50*	−12.43*	−10.1*	−11.5*
1980 presidential election	(3.4)	(3.9)	(3.4)	(3.8)
Median household income ($)		.00003		−.000023
		(.00011)		(.00011)
Percentage of adults with 4 or		2.4		1.9
more years of college		(6.3)		(6.1)
Nonwhite population percentage		.41		−.52
		(2.0)		(1.9)
Urban population percentage		−1.2		−.69
		(1.3)		(1.2)
Population per square mile		−.00032		−.00028
		(.0004)		(.00039)
Value of land and buildings per		.00004		.00003
farm acre ($)		(.000075)		(.000073)
Log likelihood	−87.3	−85.5	−90.7	−89.5
Number of observations	147	147	147	147

Note: Sample consists of all counties that had commercial hazardous waste facilities operating in 1986. Dependent variable in logit equals 1 if there is a planned net expansion in commercial hazardous waste processing capacity for 1987–92. Of the 156 counties in the full sample, 72 had positive expansion plans; among the 147 counties with no missing data, 69 had positive expansion plans. Each specification also included an intercept term and eight regional dummies. Standard errors are parentheses.

*Statistically significant at the 5 percent level.
**Statistically significant at the 10 percent level.

156 counties that in 1987 had operating commercial hazardous waste facilities by their capacity plans yields 72 counties where net planned expansion was positive and 84 counties without positive expansion plans. Table 1 shows that

waste generated, ten times the amount of value added in manufacturing, higher housing and land prices, populations with higher incomes and educations, and higher urban populations. If planned expansion at new (e.g., currently unbuilt) facilities had been included fully in the database, then the broader set of counties could be used to study expansions and new sitings. The nature of the data as a census of capacity plans for existing facilities suggests focusing on the subset of counties with facilities today and examining where expansion is planned among this set of 156 counties. Note that the set of counties examined in the logit analysis in table 2 is lower ($N = 147$) because of missing data for some variables.

the two county subgroups are extremely similar. In terms of difference-of-means tests, the counties are nearly identical in demographic factors such as household income or urban population. Capacity surpluses are smaller and hazardous waste generation is larger in counties adding capacity, but these differences are not statistically significant. The only difference of means statistically significant at the 1 percent level is related to collective action: voter turnout in those counties with positive expansion plans is .52, versus .56 in counties without net expansion plans. Table 2 presents the logit tests of facility plans, in which the dependent variable equals 1 if there is planned net expansion in commercial hazardous waste processing capacity in the county for 1987–92. Two specifications are presented, one in which a single Coasian compensation variable (*HOUSE* price) is included and a fuller specification including more Coasian variables related to factors such as compensation demand. The results in specifications 1 and 2 indicate that the higher the hazardous waste processing capacity surplus in a county (measured by the difference between the maximum processing capacity available at all facilities in the county and the utilized capacity in 1987), the less likely it is that an expansion of capacity is planned. In terms of potential demand for processing services, the higher the manufacturing value added in the county, the more likely the county is to be the site of planned expansion. Consistent with the collective action story, the higher the actual voter turnout in the county, the less likely the county is to be targeted for capacity expansion. Voter turnout in the 1980 presidential election is the only variable statistically significant in both specifications at the 1 percent level. None of the variables relating to compensation is statistically significant in either specification. A likelihood ratio test between the two specifications cannot reject the null hypotheses that the parameters on the additional variables in specification 2 are 0, and thus they provide no additional explanatory power. Specifications 3 and 4 replace the capacity variable with the total hazardous waste generated in the county (*WASTE*). Again, the voter turnout variable is statistically significant at the 1 percent level. The higher the turnout, the less likely that the county was chosen as a site for future expansion of commercial hazardous waste processing capacity. None of the variables related to compensation demands, such as income or education, is statistically significant.

In the functional forms throughout the tests presented, the Coasian variables such as income and education and the collective action variable are entered separately. This additive form implies that for a given level of the Coasian compensation variables, the higher the voter turnout, the less likely a firm will be to expand capacity in the county. This would be consistent with several explanations of how opposition is translated into effective opposition to a siting. The more politically active residents of a community are, the more likely it is that a firm may estimate that politicians will respond to opposition among constituents, that a group of residents will undertake court actions or initiate local government actions to halt sitings, or that residents and their representatives may be more skilled in negotiating fees to paid by developers, fees whose values depend in part on alternatives open to the firm in locating the facility. Each of these explanations would indicate that for a given level of opposition to a facility, a firm would expect more resistance to siting from an area with a greater potential for collective action to translate this opposition into increased location costs for the firm.

Alternative models of the collective action process might imply a multiplicative form for the Coasian compensation and collective variables, i.e., that the compensation variables should be interacted with the level of voter turnout in the county. This would be consistent with the notion that for a given amount of compensation individually demanded by residents, a firm would attach a higher probability of paying these compensation demands where individuals are politically active. The Coasian variables such as income could be interacted with voter turnout to indicate that firms view compensation demands as having a higher expected value in areas where collective action is higher. One might also use interaction terms if one believed that a given level of participation brings more results for groups with higher incomes and educations because they might be more skillful at political action. Interaction terms combining Coasian variables with voter turnout, however, generally proved to be statistically insignificant. The failure of the Coasian variables to be statistically significant in general, while voter turnout is generally significant, is consistent with emphasis on collective political action in reports on facility siting battles (see Morell and Magorian 1982; Cerrell Associates 1984).

Table 3 translates the logit coefficients into meaningful changes in probability by analyzing the impact of changes in voter turnout on expansion plans when all other variables in specification 2 in table 2 are held constant at their means. In the first row, the coefficients from the estimating equation were multiplied by the mean variable values for all 156 counties that had commercial hazardous waste facilities operating in 1986. The second figure in the first row shows the resulting probability estimate that for a county with variable values equal to those of the mean for the entire sample, the probability that a net expansion was planned in commercial capacity for 1987–92 would be .45. Consider, however, the case where the county's voter turnout rate was not the sample mean (.539) but were one standard deviation (.077) above the mean. Substituting this higher voter turnout rate into the equation results in a probability of expansion of .24; similarly, if the county had a voter turnout rate one standard deviation below the mean, its probability of being targeted for commercial expansion would rise to .68. Hence in each case a change in voter turnout by one standard deviation significantly altered the probability of expansion.

TABLE 3.   The Impact of Collective Action on the Planned Expansion of Commercial Hazardous Waste Facilties

	Probability of Planned Net Expansion of Commercial Hazardous Waste Capacity in a County, 1987–92		
	Voter Turnout = Mean − .077	Voter Turnout = Mean	Voter Turnout = Mean + .077
Variables set to means for:			
All counties with commercial capacity	.68	.45	.24
County quartile with lowest nonwhite population percentage	.60	.36	.18
County quartile with highest nonwhite population percentage	.74	.53	.30

Note: Sample consisted of all counties that had commercial hazardous waste facilities operating in 1986. The mean voter turnout in the sample was .539 and the standard deviation was .077. Probabilities estimated from specification 2 in table 2.

Further evidence on the impact of collective action is obtained by dividing the data into different county subsamples. The second row in table 3 reports the impact of voter turnout when the other variables are set at the values for the county quartile with the lowest nonwhite population percentage. The initial probability for a county with the mean variable values for this subgroup is only .36; that is, the average county with low nonwhite population percentages has about a one-in-three chance of having a planned commercial facility expansion over the years 1987–92. Yet even here, changing the voter turnout has a substantial impact on the estimated probability of expansion, lowering it to .18 for voter turnout one standard deviation above the mean and raising it to .60 for one standard deviation below the mean for that subsample. For a county with the variable values set at the mean of the county quartile with the highest nonwhite percentage, the initial probability of an expansion is .53. Changing the voter turnout rate has a substantial impact here too, lowering the expansion probability to .30 for turnout set at one standard deviation above the mean and raising it to .74 if voter turnout is set to .077 points below the subsample mean.

The division of counties by racial composition was motivated by earlier findings that "race has been a factor in the location of commercial hazardous waste facilities in the United States" (Commission on Racial Justice 1987). The specifications in table 2 show that controlling for other factors, nonwhite population percentage was not a statistically significant factor in the expansion decisions of the commercial facilities. Race is obviously correlated with other demographic factors (e.g., county nonwhite population percentage in the sample has a correlation of −.37 with median household income). The results in table 3 show that for a county with the demographic characteristics of counties with high nonwhite population percentages the estimated probability of expansion is much higher (.53) than that for a county with the characteristics of counties with smaller nonwhite populations (.36). If one looks at the actual proportion of counties targeted for expansions of hazardous waste capacity, however, the differences appear smaller across these subsamples. In the sample as a whole 46 percent of the counties with current commercial hazardous waste processing facilities are slated for additional capacity, versus 40 percent in the county quartile with the lowest nonwhite population percentage and 53 percent for the county quartile with the highest nonwhite population percentage. Given these differences, one cannot reject the hypotheses that the capacity siting proportions are the same in the counties with the highest and the lowest percentages of nonwhite populations or that the proportion of counties targeted in each of these subsamples is the same as the proportion in the other counties with operating hazardous waste processing facilities.

Tables 1 through 3 thus show that the probability of expansion varies widely among counties with currently operating commercial hazardous waste facilities; that this probability depends on the values of a collection of county variables; that after controlling for other factors, race is not a statistically significant factor in the expansion selection process; and that the potential for collective action is a sizable and statistically significant factor in the probability that a county with commercial hazardous waste facilities in operation would be targeted for planned expansion of processing capacity.[12]

Alternative Tests

One method of exploring whether the effects of the voter-turnout variable in the tests of firm expansion are truly related to collective action or due to something akin to omitted variable bias is to construct alternative tests using voter-turnout data and see if the variable behaves as predicted if it were a proxy for collective action. Consider first the expansion decisions of another category of hazardous waste facilities, generators that treat their own waste on site. Collective action may not play a large direct role in the expansion decisions of onsite generators, for at least two reasons. Those onsite generators with planned net expansions for 1987–92 on average are adding less

---

12. The magnitude of the impact of voter turnout on expansion in table 3 is similar if one uses the pared-down specification 1 from table 2 to estimate the probabilities in table 3. For a county with variable values set to the means of the counties with commercial capacity, increasing voter turnout by one standard deviation decreased the probability of expansion using specification 1 by .19 (from .47 to .28), and lowering voter turnout by one standard deviation increased the probability of expansion by .19 (from .47 to .66).

capacity than commercial processing facilities, which means they may be less opposed because their potential externalities are lower. In addition, onsite generators through their very operation provide offsetting compensation to the community in the forms of jobs and taxes, which means they would be less likely to be opposed (EPA 1979). Hence one would not expect to find onsite firms to be less likely to expand in an area if the potential for collective action were high.

In the EPA TSDR survey, there were 346 counties with onsite generators without positive net expansion plans versus 111 counties with onsite generators that had positive net expansion plans. In terms of difference-of-means tests between the two sets of counties, net expansions at onsite generators are slated for counties that are more prosperous centers of manufacturing. Manufacturing value added is twice as high in those counties with onsite expansion than those without. Income, education, urban percentage, home values, and density of population are all higher in the counties where onsite generators are expanding. Consistent with the story that collective action is not a significant determinant in the expansion decisions is that voter turnout is equal in the two subsets of counties: .54. Thus, onsite capacity processing is expanding where there is manufacturing production which generates waste.

Table 4 reports the results of logit analysis of planned expansions in counties with onsite processing facilities. The higher the manufacturing value added in the county, the more likely the county is to have a net expansion of onsite capacity planned for the 1987–92 period; in the fuller specification 2, the more urban the area, the more likely the county is to have planned onsite capacity additions. Note that the overall capacity surplus (maximum available annual processing capacity minus utilized capacity) that was negative and statistically significant in the commercial processing facilities decisions is not statistically significant, perhaps because onsite generators' decisions are driven more by their own waste production than by the overall capacity picture for all the facilities in the area. Unlike the commercial expansion process, here the voter-turnout percentage is not statistically significant in the expansion plans of onsite generators. This indicates that the negative association of voter turnout and siting is not likely to be due simply to omitted variable bias that links voting and some factor relating to industrialization. The sign of voter turnout here is positive, which may even be further indirect evidence for the role of collective action in discouraging commercial sitings. If high voter turnout results in fewer commercial facilities being sited in the area, then onsite generators in that area may be more likely to add their own capacity rather than pay higher rates to existing local facilities (partially protected by the siting barriers to entry) or pay transportation rates to more distant processors. The same results obtain if *WASTE,* county generation of hazardous waste, replaces the capacity surplus variable: onsite generators are

TABLE 4. Determinants of the Planned Expansion of Onsite Hazardous Waste Processing Facilities, 1987–92

Variable	(1)	(2)
Hazardous waste processing capacity surplus (tons)	−3.5e−11	−4.6e−9
	(1.2e−8)	(1.3e−8)
Manufacturing value added (M$)	.00025*	.00012
	(.000095)	(.000085)
Median house value ($)	4.4e−6	6.9e−6
	(9.9e−6)	(.000016)
Percentage voter turnout in 1980 presidential election	1.3	2.4
	(1.8)	(1.9)
Median household income ($)		−.000027
		(.000066)
Percentage of adults with with 4 or more years of college		−4.0
		(3.0)
Nonwhite population percentage		.21
		(1.5)
Urban population percentage		2.3*
		(.73)
Population per square mile		.00005
		(.000134)
Value of land and buildings per farm acre ($)		−.000028
		(.000034)
Log likelihood	−222.0	−215.2
Number of observations	424	424

Note: Sample consists of all counties that had onsite hazardous waste facilities operating in 1986. Dependent variable in logit equals 1 if there is a planned net expansion in onsite hazardous waste processing capacity for 1987–92. Of the 424 counties in the regression sample, 106 had positive expansion plans. Each specification also included an intercept term and eight regional dummies. Standard errors are in parentheses.

*Statistically significant at the 5 percent level.

**Statistically significant at the 10 percent level.

planning expansion where manufacturing value added is higher or in more urban areas; and collective action is neither statistically significant nor a negative factor in the expansion planning process of onsite processors.

Collective action figures not only in opposition to planned expansions but also in the shutdown of existing capacity (Lynn 1987; MacClennan 1990). The EPA's analysis (1979) of the impact of public opposition on hazardous waste facilities concluded that public pressure can result in the closure of a facility, though this may require even more political resources than blocking an initial siting. If the voter turnout variable is truly a proxy for the potential for collective action, then one might expect it to be positively associated with the decision by commercial processors to reduce their net capacity. Table 5 reports the results of examining the determinants of planned reductions in waste processing capacity by commercial facilities; of the counties with commercial facilities currently operating, 18 had planned net reductions in commercial capacity planned for 1987–92. In both model specifications, the higher the overall capacity surplus in the county (which includes capacity at commercial, limited commercial, and onsite generators), the more likely a county would be to have planned net reductions in capacity (consistent with the assumption that the higher the surplus capacity, the lower potential returns to adding or maintaining capacity). In the fuller specification, the higher the voter turnout rate, the more likely it is that net commercial capacity is slated to decrease over the 1987–92 period—evidence for the notion that greater political activity may lead to public pressure to shutdown existing commercial capacity. The higher the nonwhite population, the less likely it is that there will be a reduction in capacity (consistent with a lower compensation demand or lower political power story for minorities).

Changes in Siting Regimes

Prior to the era of publicity about groundwater contamination (highlighted at Love Canal in 1978) and the passage of the Superfund legislation in 1980, public opposition to hazardous waste facilities, while often present, was potentially less virulent and less successful in halting construction; the number of commercial facilities sited during the 1970s attest to the possibility of siting during this era. Thus if collective action became more important during the 1980s, and if voter turnout actually serves as a good proxy for the potential for collective action in an area, then the influence of voter turnout on firm location should be different if the siting process of the 1970s is compared to the expansion process of the late 1980s. This hypothesis is tested in table 6 with county census data from the 1970s and 1980s.[13] Turnout in the 1972 presidential election is used to proxy the potential for collective action during the 1970s siting regime, while turnout in the 1980 election is again used as a measure of potential political action in the 1980s.

One limitation of the data in table 6 is that information on siting of facilities in the 1970s comes from only those facilities sited that survived to be in the 1987 EPA census. Facilities that were sited and went out of business before 1987 are missing from the analysis. Second, the post-1970 sitings examined in equation 1 (a total of 133 facilities) include some facilities (11)

---

13. The census definitions of most of the variables in table 6 are the same for 1970 and 1980, except that median family income is used from the 1970 census while median household income is used from the 1980 census. Percentage of adults with four or more years of college was not in the 1970s dataset, so it is omitted from the comparable regression for 1980 "sitings." When this education variable is included in the 1980s specification in table 6, its coefficient is negative and is statistically insignificant.

TABLE 5.   Determinants of the Planned Reduction of Commercial Hazardous Waste Processing Capacity, 1987–92

Variable	(1)	(2)
Hazardous waste processing capacity surplus (tons)	2.1e–7*	2.9e–7*
	(9.8e–8)	(1.2e–7)
Manufacturing value added (M$)	−.00012	−.000027
	(.00025)	(.00035)
Median house value ($)	−.000011	.00003
	(.000025)	(.00005)
Percentage voter turnout in 1980 presidential election	5.6	10.6**
	(4.8)	(6.1)
Median household income ($)		−.000273
		(.00018)
Percentage of adults with with 4 or more years of college		−6.8
		(10.1)
Nonwhite population percentage		−6.4**
		(3.7)
Urban population percentage		3.9
		(2.6)
Population per square mile		−.000824
		(.0011)
Value of land and buildings per farm acre ($)		.00019
		(.00021)
Log likelihood	−43.2	−39.4
Number of observations	147	147

Note: Sample consists of all counties that had commercial hazardous waste facilities operating in 1986. Dependent variable in logit equals 1 if there is a planned act reduction in commercial hazardous waste processing capacity in 1987–92. Of the 147 counties in the regression sample, 18 had planned net reductions in commercial capacity. Each specification also included an intercept term and eight regional dummies. Standard errors are in parentheses.

*Statistically significant at the 5 percent level.
**Statistically significant at the 10 percent level.

sited in the early 1980s; these were included instead of selecting a cutoff date for when the stricter regulatory regime actually began. Finally, the comparisons made between the current regulatory regime and that of the 1970s involve contrasting planned expansions for 1987–92 versus actual sitings made during the 1970s. In these comparisons the focus is on which counties overall are adding capacity, so even counties that in 1986 had no commercial capacity operating but had positive net expansions planned are included among the set of counties with "sitings" (e.g., additions to capacity) for 1987–92. Unlike the previous tables, which focused on capacity expansions in a subset of counties, these specifications contrast those counties with "sitings" with all other counties in the United States.

TABLE 6. Determinants of "Siting Probabilities" Under Two Regimes,
Post-1970 Actual Sitings versus 1987–92 Planned Sitings

Variable	(1) Post-1970 Actual Sitings	(2) 1987–92 "Sitings"
Manufacturing value added (M$)	.00073*	.00041*
	(.00018)	(.00013)
Medan house value ($)	−.00006*	−6.5e−6
	(.00002)	(.000014)
Percentage voter turnout in presidential elec-	−2.10	−6.19*
tion, 1972 for (1) and 1980 for (2)	(1.52)	(1.77)
Median household income ($)	.0003*	.00016*
	(.00006)	(.00006)
Nonwhite population percentage	3.04*	1.16
	(1.07)	(1.25)
Urban population percentage	2.04*	2.15*
	(.62)	(.64)
Population per square mile	−.0003**	−.0003
	(.00015)	(.00022)
Value of land and buildings per farm acre ($)	.00003	.00005
	(.000019)	(.000045)
Log likelihood	−314.4	−282.9
Number of observations	2545	2545

Note: Dependent variable in logit (1) equals 1 if there was a commercial TSD sited after 1970 and in logit (2) equals 1 if there was planned additional capacity for 1987–92. 101 counties had sitings post-1970 versus 83 with planned capacity additions for the later period. Variable values in (1) are generally from the 1970 census, while those in (2) are generally from the 1980 census. Each regression also included an intercept and eight regional dummy variables. Standard errors are in parentheses.

*Statistically significant at the 1 percent level.
**Statistically significant at the 5 percent level.

The results in table 6 offer support for the notion that elements of the siting process have changed from the 1970s to the late 1980s. Median house value went from a negative and statistically significant impact on siting probability in the 1970s to a smaller and statistically insignificant influence in the 1980s, perhaps indicating a diminishing of purely "Coasian" concerns in the siting process as collective action became more important. Using a differenceof-means test between the coefficients on housing values in the two siting regimes, one can at the 5 percent level accept the alternative hypothesis that the impact of this Coasian variable was greater in the 1970s than in the 1980s. The impact of voter turnout on the probability of siting has also changed in the two siting processes. During the 1970s, when public opposition was not as high a barrier to entry in siting commercial hazardous waste facilities, the impact of voter turnout on siting probability was statistically insignificant. In the siting regime evident in the expansion plans for 1987–92, however, the coefficient on voter turnout is negative and statistically significant. A difference-of-means tests between the estimated voting coefficients allows one at the 5 percent level to accept the alternative hypothesis that the impact of collective action was greater in the siting regime for 1987–92. The higher the voter turnout in the county, the less likely the area was to be chosen as a site for additional commercial hazardous waste processing capacity. The results in table 6 are thus consistent with the notion both that collective action plays a significant role in the siting process in the 1980s and that this represents a change from the siting process evident in the 1970s.

6. Conclusions

This article demonstrates that commercial hazardous waste firms did take into account the potential for areas to mobilize and engage in collective action in their selection of counties in which to add capacity during the period 1987–92. Empirical models of capacity expansion were estimated that calculated the probability that a county with commercial hazardous waste processing capacity was targeted for a capacity expansion. Holding constant variables such as local capacity surpluses and the demand for waste processing or potential factors relating to compensation demands for environmental damages, the results indicate that the greater the voter turnout in a given area, the less likely that area was to be slated for expansion of commercial waste processing capacity. Evidence that the voter turnout rate is truly a proxy for collective action includes the positive association between voting rates and firm decisions to close facilities (i.e., the more politically active the community, the more likely hazardous waste facilities were to plan net reductions in capacity), the failure of voter turnout rates to be statistically significant in modeling capacity decisions at onsite generators where public opposition is not often a direct deterrent, and the increase in significance of voter turnout rates in modeling current expansion decisions versus past facility location decisions in an era (1970s) of laxer regulatory standards and lower public opposition.

In the "Coase theorem," a firm generating externalities ends up locating where, *ceteris paribus,* its social damage will be the least, because that will be where potential compensation is the least. Yet the differing degree to which groups organize to demand compensation and raise a firm's costs of choosing a particular location drives a wedge between the social costs of its externalities and the costs voiced through the political process of its site selection. In a world where collective action is variable, the firm ends up locating where the effective opposition is the least, but this may not be where the damage of its externalities is the least. If the effective externality costs estimated by a firm become a function of both the actual compensation demands of an area's residents and the probability that these will be successfully voiced through collective action, then a firm could end up locating in an area where its social costs are high but its private costs are low because of the failure of residents to oppose the siting through collective action. In terms of the Becker (1983) model of politics, the empirical results here indicate that the relative efficiency of groups in engaging in politics appears to matter a great deal in the expansion of hazardous waste capacity.

Yet one must ask why groups vary in the degree of collective action in order to assess the welfare consequences of such a siting process. This article, like Becker's model, does not provide a theory of politics which explains their variability, though unlike the latter's theory it acknowledges these differences more readily and tests to see if they have empirical applications. If the transaction costs of participating in politics explain a large degree of the differences in political participation, then one might indeed conclude that these variances in collective action give rise to sub-Pareto-optimal location decisions by the firms. Similarly, if one views the range of political participation as a function of information asymmetries in which some parties are informed about the perceived dangers of facility operation and about the mechanisms of the political process, then again there appears to be a wedge of inefficiency between the social costs of a firm's externalities and the private costs it faces. Political action also involves an individual optimizing decision in which, at least according to some theories of politics, ideological satisfaction plays a role in the decision to participate in politics. Here, in a more general framework, the tradeoff between two individual decisions—how one values the externalities generated by the facility and how one values participation in fighting it—may mean that a firm does end up choosing the Pareto-optimal location when one takes into account the perceived benefits and costs to political action by the individuals involved.

Though the normative interpretation of the results thus requires a theory of political action, the positive implications are clearer. In a world of zero transaction costs, property rights may flow to their most highly valued use. But in a truly Coasian world of transaction costs and free riders, the exercise of property rights and the consequent distribution of pollution may depend on the resolution of collective action problems.

REFERENCES

Acton, J. P. 1989. *Understanding Superfund: A Progress Report*. R-3838. Santa Monica: Rand.

Aldrich, J. H. 1990. "Turnout and Rational Choice." Working Paper no. 100, Duke University Program in Political Economy.

Andreen, W. L. 1985. "Defusing the 'Not in My Back Yard' Syndrome: An Approach to Federal Preemption of State and Local Impediments to the Siting of PCB Disposal Facilities." *North Carolina Law Review* 63:811–47.

Bacow, L. S., and Milkey, J. R. 1982. "Overcoming Local Opposition to Hazardous Waste Facilities: The Massachusetts Approach." *Harvard Environmental Law Review* 6:242–305.

Baumol, W. J., and Oates, W. E. 1988. *The Theory of Environmental Policy*. Cambridge: Cambridge University Press.

Becker, G. S. 1983. "A Theory of Competition among Pressure Groups for Political Influence." *Quarterly Journal of Economics* 98:371–400.

Carlton, D. W. 1983. "The Location and Employment Choices of New Firms: An Econometric Model with Discrete and Continuous Endogenous Variables." *Review of Economics and Statistics* 65:440–49.

Cerrell Associates. 1984. *Political Difficulties Facing Waste to Energy Conversion Plant Siting*. Report prepared for the California Waste Management Board.

Coase, R. 1960. "The Problem of Social Cost." *Journal of Law and Economics* 3:1–44.

———. 1988. *The Firm, the Market and the Law*. Chicago: University of Chicago Press.

Commission on Racial Justice, United Church of Christ. 1987. *Toxic Waste and Race in the United States*. Washington: United Church of Christ report.

Cray, C. 1991. *Waste Management Inc.: An Encyclopedia of Environmental Crimes & Other Misdeeds*. Washington, D.C.: Greenpeace.

Davis, C. E., and Lester, J. P. 1988. *Dimensions of Hazardous Waste Politics and Policy*. New York: Greenwood Press.

Deacon, R., and Shapiro, P. 1975."Private Preference for Collective Goods Revealed through Voting on Referenda." *American Economic Review* 65:943–55.

Duerksen, C. J. 1983. *Environmental Regulation of Industrial Plant Siting: How to Make It Work Better*. Washington, D.C.: Conservation Foundation.

Environmental Protection Agency (EPA). 1979. *Siting of Hazardous Waste Management Facilities and Public Opposition*. Washington, D.C.: EPA.

———. 1982. *Using Compensation and Incentives When Siting Hazardous Waste Management Facilities*. Washington, D.C.: EPA.

———. 1987. *The Hazardous Waste System*. Washington, D.C.: EPA, Office of Solid Waste and Emergency Response.

———. 1988. *Land Disposal Restrictions for First Third Scheduled Wastes*. Washington, D.C.: Office of Solid Waste and Emergency Response.

———. 1989. *Technical Reference Manual for Reporting the Current Status of Generation, Management Capacity, Imports and Exports*. Washington, D.C.: EPA, Office of Solid Waste and Emergency Response, January.

Epley Associates. 1989. *Public Relations Assessment*. Raleigh: North Carolina Low-Level Waste Management Authority.

Ferejohn, J. A., and Fiorina, M. P. 1974. "The Paradox of Not Voting: A Decision Theoretic Analysis." *American Political Science Review* 68:525–36.

Fischel, W. A. 1979. "Determinants of Voting on Environmental Quality: A Study of a New Hampshire Pulp Mill Referendum." *Journal of Environmental Economics and Management* 6:107–18.

Fortuna, R.C. 1986. *Hazardous Waste Regulation—the New Era*. New York: McGraw-Hill.

Fowler, L. L., and Shaiko, R. G. 1987. "The Grass Roots Connection: Environmental Activists and Senate Roll Calls." *American Journal of Political Science* 31:484–510.

Gardner, B. L. 1987. "Causes of U.S. Farm Commodity Programs." *Journal of Political Economy* 95:290–310.

Gergits, J. C. 1987. "Enhancing the Community's Role in Landfill Siting in Illinois." *University of Illinois Law Review* 97–129.

Goetze, D. 1982. "A Decentralized Mechanism for Siting Hazardous Waste Disposal Facilities." *Public Choice* 39:361–70.

Goldman, B. A.; Hulme, J.A.; and Johnson, C. 1986. *Hazardous Waste Management: Reducing the Risk.* Washington, D.C.: Island Press.

Gould, J. 1986. *Quality of Life in American Neighborhoods: Levels of Affluence, Toxic Waste, and Cancer Mortality in Residential Zip Code Areas.* Boulder: Westview Press and Council on Economic Priorities.

Greenberg, M. R., and Anderson, R. F. 1984. *Hazardous Waste Sites: The Credibility Gap.* New Brunswick, N.J.: Center for Urban Policy Research, Rutgers.

Harrison, D., and Rubinfeld, D. L. 1978. "Hedonic Housing Prices and the Demand for Clean Air." *Journal of Environmental Economics and Management* 5:81–102.

Harrison, D., and Stock, J. 1984."Hedonic Housing Values, Local Public Goods, and the Benefits of Hazardous Waste Cleanup." Working paper, Kennedy School of Government, Harvard University.

Hasan, N. S., and Simmons, J. R. 1989. "Local Control as a Model or Myth? The Westinghouse-Bloomington Superfund Cleanup." *Environmental Impact Assessment Review* 9:9–32.

Jorgenson, L., and Kimmel, J. J. 1988. *Environmental Citizen Suits: Confronting the Corporation.* Washington, D.C.: Bureau of National Affairs.

Kalt, J.P., and Zupan, M.A. 1984."Capture and Ideology in the Economic Theory of Politics." *American Economic Review* 74:279–300.

———. 1990."The Apparent Ideological Behavior of Legislators: Testing for Principal-agent Slack in Political Institutions." *Journal of Law and Economics* 33:103–31.

Kunreuther, H., and Kleindorfer, P. R. 1986."A Sealed-Bid Auction Mechanism for Siting Noxious Facilities." *American Economic Review* 76:295–99.

Kunreuther, H.; Kleindorfer, P. R.; Knez, P.; and Yarsick, R. 1987. "A Compensation Mechanism for Siting Noxious Facilities: Theory and Experimental Design." *Journal of Environmental Economics and Management* 14:371–83.

Lake, R. W., ed. 1987. *Resolving Locational Conflict.* New Brunswick, N.J.: Center for Urban Policy Research.

Lynn, F. M. 1987. "Citizen Involvement in Hazardous Waste Sites: Two North Carolina Success Stories." *Environmental Impact Assessment Review* 7:347–61.

McCarthy, J. 1987. *Hazardous Waste Fact Book.* Washington: Congressional Research Service.

MacClennan, P. 1990. "Jorling to Bar Waste from other States." *Buffalo News*, March 15, C1.

McConnell, V. D., and Schwab, R. M. 1990. "The Impact of Environmental Regulation on Industry Location Decisions: The Motor Vehicle Industry." *Land Economics* 66:67–81.

McFadden, D. 1973. "Conditional Logit Analysis of Qualitative Choice Behavior." In P. Zarembka, ed., *Frontiers in Econometrics*. New York: Academic Press.

Mason, G. Jr. 1989."Closure and Rejection of Waste Facilities: What Effect Has Public Pressure." *Hazardous Material Control*, July/August, 54–58.

Mitchell, R. C., and Carson, R. T. 1986. "Property Rights, Protest, and the Siting of Hazardous Waste Facilities." *American Economic Review* 76:285–90.

Moe, T. 1980. "A Calculus of Group Membership." *American Journal of Political Science* 24:593–632.

Morell, D., and Magorian, C. 1982. *Siting Hazardous Waste Facilities: Local Opposition and the Myth of Preemption*. Cambridge: Ballinger.

Mueller, D. C. 1979. *Public Choice*. Cambridge: Cambridge University Press.

O'Hare, M.; Bacow, L. S.; and Sanderson, D. 1983. *Facility Siting and Public Opposition*. New York: Van Nostrand Reinhold Company.

Office of Technology Assessment. 1983. *Technologies and Management Strategies for Hazardous Waste Control*. Washington D.C.: GPO.

———. 1985. *Superfund Strategy*. Washington, D.C.: GPO.

———. 1986. *Serious Reduction of Hazardous Waste for Pollution Prevention and Industrial Efficiency*. Washington, D.C.: GPO.

Olson, M. 1965. *The Logic of Collective Action*. Cambridge: Harvard University Press.

Pashigian, B. P. 1985. "Environmental Regulation: Whose Self-interests Are Being Protected?" *Economic Inquiry* 23:551–84.

Pearl, D. 1991. "Neighborhoods Resist Recycling Plants." *Wall Street Journal*, October 14, B1.

Percival, R.; Miller, A.; Schroeder, C.; and Leape, J. 1992. *Environmental Regulation: Law, Science, and Policy*. Boston: Little Brown.

Riker, W. H., and Ordeshook, P. C. 1968. "A Theory of the Calculus of Voting." *American Political Science Review* 62:25–42.

Roper Organization. 1990. *The Environment: Public Attitudes and Individual Behavior*. Report commissioned by S.C. Johnson and Sons, July.

Schneider, K. 1990. "Safety Fears Prompt Plants to Buy Out Neighbors." *New York Times*, November 28, A1.

Schwab, S. 1989."Coase Defends Coase: Why Lawyers Listen and Economists Do Not." *Michigan Law Review* 87:1171–98.

Smith, V. K., and Desvousges, W. H. 1986a. "The Value of Avoiding a LULU: Hazardous Waste Disposal Sites." *Review of Economics and Statistics* 68:293–99.

———. 1986b. "Asymmetries in the Valuation of Risk and the Siting of Hazardous Waste Disposal Facilities." *American Economic Review* 76:291–94.

Stever, D. W. 1988. "Remedies for Hazardous or Toxic Substance-related Personal Injuries: A Discussion of the Usefulness of Regulatory Standards." *Houston Law Review* 25:801–15.

Sullivan, A. M. 1987. "Policy Options for Toxics Disposal: Laissez-Faire, Subsidization, and Enforcement." *Journal of Environmental Economics and Management* 14:58–71.

———. 1990. "Victim Compensation Revisited: Efficiency versus Equity in the Siting of Noxious Facilities." *Journal of Public Economics* 41:211–25.

Verba, S., and Nie, N. 1972. *Participation in America*. New York: Harper and Row.

Walsh, E. J., and Warland, R. H. 1983. "Social Movement Involvement in the Wake of a Nuclear Accident: Activists and Free Riders in the TMI Area." *American Sociological Review* 48:764–80.

White, M. J., and Wittman, D. 1981. "Optimal Spatial Location Under Pollution: Liability Rules and Zoning." *Journal of Legal Studies* 10:249–68.

Wolfinger, R. E., and Rosenstone, S. J. 1980. *Who Votes?* New Haven: Yale University Press.

Part 5
Environmental Treaties

CHAPTER 10

Private Provision of a Public Good: A Case Study

Lars P. Feld, Werner W. Pommerehne, and Albert Hart

1. Introduction

It is more or less common knowledge in public finance literature that public goods, i.e., goods and services whose benefits are nonexcludable, cannot be provided within a private market setting, since each citizen will occupy a free-rider position, i.e., one hopes that others will bear the cost of providing the public good while personally enjoying the respective benefit (Samuelson 1954, 389; Musgrave and Musgrave 1984, 50).

With respect to the free-rider assumption in its strong version (i.e., the calculus of each citizen is as just outlined), however, there exist a number of experiments suggesting that this hypothesis cannot be maintained.[1] In all studies where the conditions were investigated under which free-riding behavior may occur, the strong free-rider hypothesis can be rejected, since participants revealed voluntarily between 50 percent and 70 percent of the assumed true willingness to pay for the good under concern. Nevertheless, these results were achieved mostly in experimental settings (often in the form of one-shot experiments). Their hypothetical nature could cast some doubt as to their reliability.[2]

This article is a revised and expanded version of a paper that appeared in *Kyklos* 1994. Werner W. Pommerehne died suddenly, unexpectedly, and untimely on October 8, 1994. The revision of the paper is based on commonly discussed notes. The authors would like to acknowledge a research grant from the Deutsche Forschungsgemeinschaft (DFG) as well as the substantial contribution of Joachim Nick (Federal Ministry of the Environment, Bonn) in developing and administering the survey. They would also like to thank Jürgen Backhaus, Roger D. Congleton, Bruno S. Frey, Anselm U. Roemer, and Georg Seeck for useful comments. Helpful research assistance was provided by Cornelia Bitter, Ulrike Krausenbaum, and Bodo G. Schirra.

1. The seminal experiment is due to Bohm 1972. Further evidence with modified scenarios can be found, e.g., in Smith 1979; Brubaker 1982, 1984; Isaac, McCue, and Plott 1985; Andreoni 1988; Palfrey and Rosenthal 1988; Van de Kragt, Dawes, and Orbell 1988; Mestelman and Feeny 1988; and Bagnoli and McKee 1991.

2. However, in a few cases, as in Schneider and Pommerehne 1981, pretending a real situation also led to the rejection of the strong free-rider hypothesis.

A second sort of evidence against the strong version of the free-rider hypothesis can be derived from (more or less unique) events in reality. Weisbrod 1975, as well as Weisbrod and Dominguez 1986, points out that the voluntary support of nonprofit organizations in the United States, which provide goods and services for the poor and the aged, is compatible with the basic axioms of the standard neoclassical literature only under very restrictive assumptions.[3]

A third type of evidence against the strong free-rider hypothesis can be found in various events mostly observed in developing countries.[4] Ostrom 1990 analyzed a number of cases where individuals were using common property resources over a long period of time.[5] In contrast to the standard case, no overuse occurred, i.e., under specific institutional settings individuals did not adopt a complete free-rider position and did not increase continuously their use of the resource.[6]

As a fourth reason for privately contributing to the provision of a public good, game theoretic arguments were brought forward, e.g., by Gaitsgory and Nitzan (1993). In a dynamic setting, it can be shown that a dominant strategy for individuals may very well consist in acting cooperatively.[7] This draws attention to the fact that in conventional analysis of free riding behavior, the situation is modeled as a one-shot game. This, in fact, might produce misleading conclusions in that the actual extent of free riding will be constantly overestimated.

The basic difficulty in testing the free-rider hypothesis consists in the fact that the test situation is usually not appropriate. What is required is an adequate "Coase-like situation," where the assignment of property rights was carried out in the past and in which the citizens who do not possess the property right have no other option than to provide the public good voluntarily. Such situations are very rare since in reality people can pursue a number of strategies rather than stating their private willingness to contribute to the provision. For instance, citizens may exercise voice, i.e., try to pressure the politicians directly to provide the public good in question.[8] With respect to public bads, e.g., negative environmental externalities, they may

---

3. Similar evidence is provided by Schiff 1990. For a presentation and discussion of the underlying idea of participation altruism see Margolis 1982.

4. See, e.g., Bolnick 1976; Kikuchi, Dozina Jr., and Hayami 1978; and Wilson 1992; yet similar evidence can also be found in the America of the early times (see, e.g., Klein 1990).

5. Ostrom's examples of common property resources cover, e.g., woods in Japan and Switzerland, irrigation projects in the Philippines and Spain, and groundwater use in California.

6. See Hardin's (1968) famous essay "The Tragedy of the Commons" for a vivid description of the standard overuse problem.

7. The original idea was developed by Axelrod (1984).

8. A particularly attractive form of this strategy may exist in federal systems, where politicians of the central government may be pressed to provide a local public good, since in this case the citizens from all other states bear most of the respective financial burden.

seek out the courts to forbid (or at least reduce) the activity creating the externality.[9] In none of these cases does a Coase-like situation prevail and each attempt to test the free-rider hypothesis has to deal with the problem of handling these disturbances.[10]

In the following sections we report on a test of the free-rider hypothesis in a natural setting.[11] Section 2 sketches briefly the specific situation in which the voluntary provision of a public good has been attempted. Section 3 empirically explores the reasons for and the size of the individual contributions based on a data set of interviewed citizens, while section 4 provides some sensitivity analysis on results obtained in section 3. Section 5 contains some concluding remarks.

2. The Kleinblittersdorf Case

The community of Kleinblittersdorf, which is located in the German state of Saarland and directly on the French border, was faced with the following situation. In February 1987, the French sister community Grosbliederstroff, situated to the west just across the River Saar, planned to build and operate a waste incinerator in a derelict and contaminated area within the community. Since west winds strongly predominate in this region, the citizens of Kleinblittersdorf felt threatened by the pollution stemming from the waste incinerator. Public meetings were held in order to inform the citizens about the degree of pollution and the potential toxicity of the pollutants. Besides the health aspect, expected losses with respect to real estate values were widely discussed. At these meetings and on other occasions, representatives as well as citizens were engaged in several kinds of protest actions, such as writing petitions and collecting signatures against the incinerator. For instance, up to 1988, a total of 16,000 legal objections was presented to the prefect in Metz (the capital of the French Département Moselle).

Furthermore, the mayor of Kleinblittersdorf, Robert Jeanrond, successfully tried to induce the government of Saarland to stop its support for the French incinerator. The government, in contrast to its earlier position,[12] announced that the state of Saarland would not deliver any waste to the French incinerator. However, due to the fact that Saarland lacked the capacity to treat waste and due to the institutional and factual difficulties associated with

---

9. This is usually the most common procedure in the case of NIMBY goods, i.e., goods with locally concentrated negative benefits and widespread regional positive benefits.

10. Laboratory experiments and contingent valuation studies also have to deal with this problem, since some respondents may not accept the stated scenario.

11. See also Pommerehne, Feld, and Hart 1994.

12. Originally the government of Saarland had regarded the plan to build an incinerator positively, as it offered an opportunity to get rid of parts of the Saarland's waste.

building an incinerator in the state of Saarland, it remained an open question whether this announcement constituted a credible commitment. With respect to other German states that were also interested in exporting waste to Grosbliederstroff, Jeanrond was even less successful, since some states declared in a rather nonbinding manner that they would not deliver waste to Grosbliederstroff while others even declined to go along with this proposal. Furthermore, even if a credible commitment had been achieved with the consequence that no German waste would be delivered to the French incinerator, the net effect for the citizens of Kleinblittersdorf would have been ambiguous. Though the quantity of treated waste would have been reduced, the health damage and other inconveniences could have been as large as or even larger than in the case of additional waste imports from Germany. The reason was that the French operator had threatened in the case of a German waste embargo to construct the incinerator in accordance with the much less stringent French environmental standards, whereas he had offered to comply with the German standards in the case of German exports.

Besides these activities, Jeanrond urged the government of Saarland to negotiate with the French regional government (Département Moselle) as well as directly with the central government in Paris in order to persuade them to abstain from building the waste incinerator. Later, Jeanrond contacted French politicians for the same purpose. However, it turned out that the chances of success in preventing the waste incinerator through legal action and political pressure seemed rather low. In such a (Coase-like) situation no other options were left but to negotiate directly with the French community of Grosbliederstroff in order to achieve a mutually beneficial solution.

In June 1987, Jeanrond came up with the plan to buy the derelict and polluted land from the French municipality and to find an investor who would operate a less pollution-intensive plant. After substantial evaluation of a number of candidates, a supplier of automobile parts was selected as the most convenient candidate,[13] though he demanded as precondition a cleanup of the land at the expenses of Kleinblittersdorf and substantial financial aid to realize his investment plans. Jeanrond persuaded the French community of Grosbliederstroff, which also perceived benefits from this less pollution-intensive investment project, to sell the area to him for the symbolic price of one franc. Following this, public meetings were held in Kleinblittersdorf to inform the citizens about the plan, which required the total sum of DM 3.5m to be collected.[14] After explaining carefully the expected (relative) benefits of

---

13. Besides the health aspect, this investor offered to create 100 additional jobs in the short run, and 300 in the longer run, compared to the employment effects of the incinerator project.

14. This is the nominal amount required. However, those citizens who wanted to give a contribution had the choice whether they wanted to provide an out-of-pocket donation, a subsi-

the new plant with respect to health indicators as well as real estate losses, Jeanrond suggested that each citizen should contribute DM 800 or 10 percent of the expected loss of the real estate values that would occur if the incinerator was built. The appeal was repeated several times in the newspapers and at various meetings. In order to organize the collection and the administration of the donated money a club was established. It received financial contributions from June 1988 on.

One might raise the question why the cost of preventing the construction of the waste incinerator was not covered by a variation of the local budget. Why, for example, weren't the rates of local business and real estate taxes raised (i.e., the only taxes that are assigned to the local level)? At a first glance, one might think there seemed to be at least some leeway in raising the rates of both taxes as compared to the local community average (on the statewide level). This argument, however, is not valid since the project itself would have required a total of DM 3.5m, a sum equivalent to no less than 20 percent of the overall budget of Kleinblittersdorf. Therefore, it would have been rather unlikely that this huge amount could have been raised simply by increasing local tax rates in the short run. Moreover, in comparison to the local business and real estate taxes of the neighboring municipalities, Kleinblittersdorf's tax rates were quite high, providing another reason not to increase the local tax burden. Furthermore, it was not possible to increase the municipality's deficit, due to the strong constraints imposed by the "Kommunalaufsichtsbehoerde," i.e., a supervising state authority on local budget affairs. Cutting down on current local expenditures was not possible either, because the bulk of them consisted in compensation of local public employees and in other spending items (like social and housing assistance) which cannot be reduced in the short run. Finally, local government investment expenditures could not be lowered because they are, in general, earmarked for long-run purposes and, in addition, subject to strong control by the Kommunalaufsichtsbehoerde.

Thus the stage was set for a real Coase-like bidding game where the contributions to the project had to be negotiated individually. In fact, individual bidding took place to a considerable extent. By July 1990, a total of DM 3,525,955 had been collected, exceeding the required amount by some DM 25,000.

In the summer of 1991, a university research team surveyed a sample of randomly chosen inhabitants of the community of Kleinblittersdorf. The main target of the survey was to get an idea of factors that determined the amount

---

dized loan (with an interest rate lower than the market rate), or a subsidized loan in combination with a guarantee. When we use the term contribution in the following we always refer to the present value of the financial burden that results from each form of contribution.

people actually contributed, in order to elicit some underlying motives as well as constraints.

3. Estimating a Contribution Function

3.1. Basic Estimation

The survey was based on a random sample of households in the municipality of Kleinblittersdorf. It was attempted to question a representative member (above the age of 18) of each selected household.[15] A letter was sent to 400 randomly chosen households, inviting them to participate in an interview. Two hundred one completed interviews were yielded, which corresponds to about 3.5 percent of the households.

In order to estimate a contribution function, i.e., to explain the individual financial contributions toward creation of a beneficial alternative to the incinerator, the individual (household) contributions (and their potential determinants) were captured through self-reports of the respondents. In the following, $LCON$ presents the logarithmic value of the contributions.[16] In the simplest case, these may only depend on the net household income (which is represented by the logarithm of the household income per year, $LHHINC$). In this case, it is assumed that all individuals do not only have the same utility function but also expect the same damage. The result of the respective OLS-estimation is as follows.[17]

$$LCON = -42.334 + 4.146* \quad LHHINC,$$
$$(2.106)$$

$$R^2 = 0.022; \; \bar{R}^2 = 0.017; \; \text{d.f.} = 199; \; F = 4.435. \tag{1}$$

The coefficient on household income is positive and statistically significant. However, the R^2 is very low, even for cross-sectional estimation standards, suggesting that there are other major determinants, which may be much

---

15. Thus, following Becker 1974, our basic economic entity is the household.

16. Zero contributions were slightly corrected upwards by an infinitesimal amount in order to keep the estimation approach tractable. The same procedure was applied to any other log-term as well.

17. In the following, two asterisks indicate statistical significance at the 99 percent confidence level, one asterisk (in parentheses) indicates statistical significance at the 95 (90) percent confidence level. The numbers in parentheses represent the t-value of the respective coefficient. The term R^2 ($\bar{R}^2$) corresponds to the coefficient of determination (corrected by the degrees of freedom); d.f. provides the degrees of freedom, and the F-value indicates the statistical significance of the estimate in total.

more important than household income. Of course, the assumption that every citizen expects the same damage is not realistic: Citizens who are living farther away from the incinerator may expect less damage and, regarding their personal circumstances, they will not be affected in the same way.

Consequently, people were asked to state the subjectively perceived personal damage that would arise if the incinerator was built. Two broad categories of damage can be distinguished: health effects and financial losses associated with the expected decline of real estate prices. With respect to health damage, participants were asked to assess their actual personal health status on a scale from zero (for a very bad state) to six (for an excellent state).[18] They were also asked what health status they were expecting if the incinerator was erected. The variable *HEALTHLOSS* reflects the difference between the two ratings, i.e., the deterioration of the health status the citizens associate with the incinerator. Consequently, the variable takes on values from zero to six with zero meaning no deterioration and six meaning very strong deterioration. Participants who owned or expected to inherit real estate were also asked to provide an estimate of expected loss of real estate values (in percent). Since the (historic) values of their real estate were also recorded,[19] the variable *LRESLOSS* reflects the natural logarithm of the expected financial loss associated with the decline in real estate values.[20] For both loss variables a positive influence on the individual contributions is expected.

Moreover, the interpretation of the expected damage is not independent from the stated levels of health and of real estate values: It makes a difference whether a marginal deterioration of health due to the construction of the incinerator occurs when a person is in a state of good health, or whether the same health deterioration occurs in a state of bad health. Therefore, we would expect that a participant with an initial state of excellent health contributes less, ceteris paribus, than one with an initial state of bad health. The same logic applies to real estate values.[21] The initial state of health is captured by

---

18. This question was directed at actual health status (with no incinerator in operation). In the following it is assumed that this health status would not deteriorate through the operation of the plant of the automobile supplier.

19. We got the impression that most respondents stated the assessed value of their real estates rather than the (potential) market value. This is not surprising, since the first is a well-known component of the yearly tax declaration, while the second may be unknown unless respondents try to sell the respective properties.

20. Theoretically, the stated real estate values could already reflect the losses associated with the incinerator. However, in that case we would expect the values to decline as the distance to the incinerator increases. The corresponding univariate regression shows no significant impact of the distance neither on the real estate values (t-value -0.142) nor on the *expected* financial loss of real estate values (t-value -1.353), whereas it has a significant impact in the case of *HEALTHLOSS* (t-value -1.951).

21. The idea of the importance of relative changes was put forward by Kahneman and Tversky (1979).

the variable *HEALTHSTATE* with values ranging from zero to six, while real estate values are used in the form of logterms (*LRESTATE*).

Since, according to Becker's (1974) theory of household altruism, parents incorporate their children's well-being in their own utility function, we might expect that especially a long-run disamenity like a waste incinerator would strongly decrease the utility level of households with children.[22] Since only the fact that people have children at all matters in this respect, we incorporate a children-dummy (*DYCHILD*) in our estimation setup. As a consequence, we expect households with children to contribute relatively larger amounts, ceteris paribus. With these additional variables our estimation reads as follows;

$$LCON = -22.094 + 1.726 \; LHHINC + 1.918** \; HEALTHLOSS$$
$$(0.897) \qquad\qquad (2.636)$$

$$- \; 1.566* \; HEALTHSTATE + 3.764^{(*)} \; DYCHILD$$
$$(-2.122) \qquad\qquad\qquad (1.651)$$

$$+ \; 0.225* \; LRESLOSS + \; 0.108 \; LRESTATE,$$
$$(2.404) \qquad\qquad\quad (1.037)$$

$$R^2 = 0.185; \; \bar{R}^2 = 0.160; \; \text{d.f.} = 194; \; F = 7.360. \tag{2}$$

The two variables related to the expected decline in health and estate value have a positive sign and are strongly significant, though only the case of the respective initial level of health (*HEALTHSTATE*) has the expected negative and statistically significant coefficient. These results confirm our a priori expectations that the respondents' contributions are the higher the higher the expected loss that is associated with the incinerator. Yet, the marginal impact is significantly lower the better the initial health state. *DYCHILD* has the expected positive sign and is statistically significant at the 90 percent confidence level. Household income and the real estate value which are both insignificant (in this and all further regressions) are dropped. When we redo the estimation without these two variables we obtain

$$LCON = -8.000 + 1.831* \; HEALTHLOSS - 1.439^{(*)} \; HEALTHSTATE$$
$$(2.532) \qquad\qquad\qquad (-1.990)$$

$$+ \; 4.336^{(*)} \; DYCHILD + 0.309** \; LRESLOSS,$$
$$(1.896) \qquad\qquad\quad (4.931)$$

$$R^2 = 0.178; \; \bar{R}^2 = 0.161; \; \text{d.f.} = 196; \; F = 10.600. \tag{2'}$$

---

22. This insight is also used, e.g., by Barro (1974) in his theory of Ricardian equivalence.

Although the R^2 has risen considerably compared with the naive estimation at the outset, it nevertheless remains at a rather low level. Therefore, we look at further potential factors influencing the contribution decision.

3.2. Information Aspects

The conventional theory of household behavior relies on the premise of perfect information. In reality, however, this assumption is violated in many cases. Different degrees of knowledge of the environment should have a crucial influence on behavior resulting from optimization under constraints.[23] Individual choice, in other words, can be thought of as being state-dependent where the state in this case refers to the degree of information.

To the citizens of Kleinblittersdorf it soon became clear that any effort to prevent the construction by exerting voice was doomed to failure. Taking this into account, we can expect that informed citizens realized that the only way to prevent the construction of the incinerator consisted in private contributions. Therefore, the contributions of those individuals who stated that they had a relatively high degree of information should be relatively larger than the others', ceteris paribus.

On June 1, 1988, the mayor of Kleinblittersdorf invited the population to a general meeting where he notified the participants of the events which had been going on so far. Basically, three steps had been taken up to this point: (1) the establishment of two clubs to coordinate the collection and the administration of the contributions; (2) calculation and elicitation of the overall amount required to prevent the incinerator; (3) the information that real estate values were down already by some 30 percent. Finally, the mayor also presented his calculation that, in order to raise enough money to prevent the construction of the incinerator, the necessary contribution of each citizen would be DM 800 on the average. Those people who did not attend that meeting were asked about their knowledge of these four events. From their answers, an information variable is constructed reflecting the sum of the corresponding four information dummies. Those individuals who attended the 1988 meeting are assumed to have processed those four informational aspects. Thus, we obtain the ordinal variable *INFOSCALE* with values ranging from zero (none of the events was known) to four (all events were known). With this additional variable we get the following estimation results

$$LCON = -23.048 + 1.245^{(*)} HEALTHLOSS$$
$$(1.747)$$

---

23. See the pioneering work by Stigler (1961).

$$- 1.269^{(*)} \; HEALTHSTATE + 4.139^{(*)} \; DYCHILD$$
$$(-1.816) \qquad\qquad (1.877)$$

$$+ 0.268^{**} \; LRESLOSS + 4.496^{**} \; INFOSCALE$$
$$(4.373) \qquad\qquad (3.976)$$

$$R^2 = 0.240; \; \bar{R}^2 = 0.220; \; \text{d.f.} = 195; \; F = 12.282. \qquad (3)$$

The coefficient of the information variable shows the expected sign and is strongly significant. This additional variable improves the goodness of the estimation by about six percentage points. Compared to equation (2') the coefficients of the further variables only change somewhat, indicating the robustness of the relationship.

3.3. Coordinating Efforts

The city of Kleinblittersdorf is inhabitated by some 16,000 people. Given so large a group, the question can be raised whether individual negotiations could bring about the prevention of the incinerator without external coordination. And, indeed, it turned out that mayor Jeanrond played a significant role in organizing collection efforts. Toward this end he engaged in two major activities right at the beginning of the donation drive, both in the form of appeals to the citizenry. The first consisted in challenging the inhabitants of Kleinblittersdorf to contribute DM 800 each. As a second activity, Jeanrond published a letter in the local newspaper where he appealed to real estate owners to donate 10 percent of the assessed (30 percent) downfall of real estate values.

The effect of each appeal was asked for in the interview. In the case a participant stated that he was guided by the mayor's appeal to contribute DM 800, the ratio of the respective household income to the amount of DM 800 is calculated (*LRINC800*) indicating the ease or the degree of liquidity with which this sum can be contributed.[24] Otherwise, *LRINC800* takes on the value of zero. In case a respondent reported having been influenced by the second appeal, this is processed by calculating the natural logarithm of the monetary amount corresponding to the requested 3 percent (0.1×0.3) of the respective real estate value (*LREQ3%*). Otherwise, *LREQ3%* is set to zero. Taking into account the impact of these coordinating activities, we obtain the following results

---

24. Actually, *LRINC800* is calculated as the natural log of the ratio, augmented by one for ease of calculation.

$$LCON = -22.241 + 1.472* \; HEALTHLOSS$$
$$(2.288)$$

$$- 1.440* \; HEALTHSTATE + 3.121 \; DYCHILD$$
$$(-2.292) \qquad\qquad\qquad (1.561)$$

$$+ 0.175** \; LRESLOSS + 3.477** \; INFOSCALE$$
$$(3.084) \qquad\qquad\qquad (3.387)$$

$$+ 8.342** \; LRINC800 + 0.684** \; LREQ3\%$$
$$(6.140) \qquad\qquad\qquad (2.582)$$

$$R^2 = 0.393; \; \bar{R}^2 = 0.370; \; d.f. = 193; \; F = 17.811. \qquad (4)$$

The comparison of equation (3) and equation (4) reveals that the coefficients of most variables used so far lose somewhat in quantitative importance and statistical significance when proxies for the coordination efforts of the mayor are included. The coefficients of both appeal variables have the expected positive sign and are statistically significant at the 99 percent confidence level. This as well as the fact that R^2 rises from 24 to 39 percent indicates that the coordination efforts play a nonnegligible role in explaining the final success of the contribution mechanism.

3.4. Embeddedness and Related Influences

If we abstain from a purely public good view of the Kleinblittersdorf case, we might very well realize that arguments originating from interest group theory turned out to be relevant in that situation. As Olson (1965) points out, selective incentives which are set by individual actors could be a crucial factor in the provision of a collective good. Influence in that direction might have been exerted by the invitation to attend the general meeting of June 1, 1988. Those who participated might have been, according to Klein's (1990) interpretation of Olson's idea, integrated into a network of people willing to prevent the construction of the incinerator. This special engagement can be expected to exert a positive influence of its own on the willingness to contribute, due to some kind of (self-justifying) embeddedness.

Engagement in this kind of activity is captured by the dummy variable *DYMEET*. The extended estimation results are as follows:

$$LCON = -21.036 + 1.194^{(*)} \; HEALTHLOSS$$
$$(1.830)$$

$$- 1.304* \; HEALTHSTATE \; + \; 3.600^{(*)} \; DYCHILD$$
$$(-2.081) \qquad\qquad\qquad (1.803)$$

$$+ \; 0.168** \; LRESLOSS \; + \; 2.695* \; INFOSCALE$$
$$(2.984) \qquad\qquad\qquad (2.480)$$

$$+ \; 7.845** \; LRINC800 \; + \; 0.634* \; LREQ3\%$$
$$(5.731) \qquad\qquad\qquad (2.405)$$

$$+ \; 4.400* \; DYMEET,$$
$$(2.056)$$

$$R^2 = 0.406; \; \bar{R}^2 = 0.381; \; d.f. = 192; \; F = 16.374. \qquad\qquad (5)$$

The congregation dummy exhibits the expected sign and is significant at the 95 percent confidence level.[25] However, the goodness of the fit increases only slightly.

Embeddedness based on economic motives might be too narrow a concept when approaching contribution activities on a community level. Collective activities, especially when they are taking place in a small municipality, can be expected to be backed up by the network of interactions among inhabitants who know each other very well. In light of this, we may conjecture that the longer an individual has been a member of this local community, the larger the pressure on him/her to participate in collective actions. This "social force" affects his/her decisions via communication with other inhabitants and can be understood as a particular restriction in his/her decision set. The restriction should be especially strong if a person has many friends in the local community. In this case, he/she is more likely to listen to suggestions made by other people.

While this constraint can be categorized as noneconomic, there may be additional economic constraints on individuals' decisions as well. For instance, the more children are living in a household, the more difficult it gets to exercise the option to move to another community (especially if the children attend different schools). Setup costs, so to speak, become particularly high when such a family settles in another municipality.[26]

---

25. Since DYMEET is also incorporated in INFOSCALE, we would expect some collinearity between both variables; however, the corrected Pearson's contingency coefficient is only about 0.65.

26. An alternative explanation is provided by Hirschman (1982, 79ff.). He emphasizes that sunk costs do determine people's decisions in reality. Following this argument, the number of children could also be understood as an indicator of the relevance of sunk costs (especially if people act retrospectively).

Social embeddedness is measured by the length of time an individual has been living in Kleinblittersdorf. On the other hand, the number of children in the respective household serves as a proxy for private motives of engaging in local community activities for the aforementioned reasons. Social pressure, coded as *LLIVTIME,* is expected to raise the individual contribution while the same should be true for the number of children, the relevant variable being denoted *LNUMCHILD* in our enlarged estimation set-up. The results read as follows.

$$LCON = -24.079 + 1.129^{(*)} \ HEALTHLOSS$$
$$(1.734)$$

$$- \ 1.286* \ HEALTHSTATE \ - \ 2.924 \ DYCHILD$$
$$(-2.054) \qquad\qquad (-0.489)$$

$$+ \ 0.144* \ LRESLOSS \ + \ 2.453* \ INFOSCALE$$
$$(2.446) \qquad\qquad (2.214)$$

$$+ \ 7.713** \ LRINC800 \ + \ 0.662* \ LREQ3\%$$
$$(5.643) \qquad\qquad (2.503)$$

$$+ \ 4.425* \ DYMEET \ + \ 1.239 \ LLIVTIME$$
$$(2.075) \qquad\qquad (1.281)$$

$$+ \ 7.490 \ LNUMCHILD$$
$$(1.250)$$

$$R^2 = 0.416; \ \bar{R}^2 = 0.385; \ \text{d.f.} = 190; \ F = 13.536. \qquad\qquad (6)$$

Both additionally included variables have the expected positive sign, but none of them is statistically significant. Moreover, the coefficient of the children dummy, being positive and significant in equation (5), becomes insignificant in equation (6). A look at the corrected Pearson's contingency coefficient[27] indicates that there is a high degree of association of *DYCHILD* and *LNUMCHILD.* Therefore, equation (6) is redone, leaving out the most insignificant children dummy

$$LCON = -24.219 + 1.145^{(*)} \ HEALTHLOSS$$
$$(1.765)$$

$$- \ 1.294* \ HEALTHSTATE \ + \ 0.143* \ LRESLOSS$$
$$(-2.072) \qquad\qquad (2.436)$$

---

27. This coefficient has a value of 99 percent.

$$+ \ 2.411* \ INFOSCALE \ + \ 7.722** \ LRINC800$$
$$(2.187) \qquad\qquad\qquad (5.661)$$

$$+ \ 0.671* \ LREQ3\% \ + \ 4.484* \ DYMEET$$
$$(2.544) \qquad\qquad\quad (2.110)$$

$$+ \ 1.295 \ LLIVTIME \ + \ 4.734* \ LNUMCHILD$$
$$(1.352) \qquad\qquad\quad (2.339)$$

$$R^2 = 0.415; \ \bar{R}^2 = 0.388; \ \mathrm{d.f.} = 191; \ F = 15.074. \qquad (6')$$

As expected, the number of children now becomes significant in equation (6') and the statistical significance of the whole set of variables included (F-value) improves compared to equation (6). Moreover, the impact of all other variables remains about the same as in equation (5). Since the t-value of the coefficient of *LLIVTIME* is considerably larger than one, we decide to keep this variable in the setups which follow.[28]

3.5. Free Riding, Protest, and Exit

A number of participants in our interview reported a zero contribution. Two quite different reasons could account for abstaining from contributions. First, of course, the well-known free-rider attitude comes to mind. According to standard economic theory, the conjecture that others will participate in providing a public good should decrease the amount an individual is willing to contribute.[29] The reason for this is to be seen in purely egoistic motives and, therefore, in the unwillingness to act independently of other peoples' behavior.

Judging from this, we would expect that, when it comes to voluntary donations for whatever cause, a person with a free-rider attitude is consistently less willing to engage in those activities. Hence, a good proxy to capture free-riding motives consists in a variable which provides evidence of the general donation behavior of an individual in the past. By asking people's former overall contributions to a number of largescale projects (such as environmental improvement, aid to developing countries, etc.) we, in fact, capture the degree to which a free-rider attitude was prevalent in the subjects of our sample. The amount of donations is related to the corresponding person's household income in order to get a measure of the degree of voluntary partici-

---

28. Unfortunatly, the influence exerted by friends turned out to be very badly measured in the interview, mainly due to the fact that the respective question was not appropriately standardized. Therefore, we make no use of the respective answers.

29. See Margolis's (1982, 19ff.) famous radio station example.

pation.[30] Subtracting this ratio from its maximum provides a standardized continuous proxy for a free-rider attitude. The use of a continuous scale, *FREESCALE*, takes account of the fact that partial free-riding behavior is possible. *FREESCALE* is expected to have a negative impact on contributions, which is confirmed by equation (7).

$$LCON = -14.885 + 0.980 \; HEALTHLOSS$$
$$(1.514)$$

$$- \; 1.089^{(*)} \; HEALTHSTATE + 0.141* \; LRESLOSS$$
$$(-1.741) \qquad\qquad (2.441)$$

$$+ \; 2.170* \; INFOSCALE + 7.778** \; LRINC800$$
$$(1.978) \qquad\qquad (5.758)$$

$$+ \; 0.672* \; LREQ3\% + 5.011* \; DYMEET$$
$$(2.573) \qquad\qquad (2.366)$$

$$+ \; 1.272 \; LLIVTIME + 4.926* \; LNUMCHILD$$
$$(1.340) \qquad\qquad (2.456)$$

$$- \; 10.553* \; FREESCALE$$
$$(-2.184)$$

$$R^2 = 0.430; \; \bar{R}^2 = 0.400; \; \text{d.f.} = 190; \; F = 14.312. \qquad\qquad (7)$$

The structure of the estimation results is not affected to a large extent by the inclusion of *FREESCALE*. An exception are both health variables, whose marginal impact and significance decline somewhat.

A second determinant of noncontribution may consist in the fact that some people might be ardent opponents of the mechanism of privately contributing, at least in our context. Thus, there may have been people whose theoretical willingness to contribute was definitely positive, but who refused to participate, out of fundamental opposition to the proposed type of problem solution.[31] This attitude must be strictly separated from free-rider motives, because it is not egoistic reasons which cause those people to stay away.

Since this specific attitude is, in contrast to a free-rider attitude, incompatible with participating in the contribution campaign at all, we resorted to

30. The mean of this ratio is 0.063 (s.d. = 0.092) overall and 0.075 (s.d. = 0.096) only for those who donated a positive amount. The maximum of 0.5 occurred in only four cases, suggesting quite a high degree of reliability of the answers.

31. Ng (1988) terms this phenomenon "procedural preference"; see e.g., also Okun 1975, 47.

asking only those who did not contribute for potential reasons of protest. From the offered items we select three that we think are valid reasons for being against the private provision.[32] The variable PROTEST consists in the weighted sum of those three reasons for the corresponding noncontributor's decision. It ranges from zero (none of the reasons was relevant) to nine (the three reasons are the most relevant ones among a larger number of potential candidates). It goes without saying that the expected sign of the respective coefficient should be negative. Inclusion of this variable in our estimation setup yields equation (8).

$$LCON = -11.221 + 0.626 \ HEALTHLOSS$$
$$(1.088)$$

$$- \ 0.700 \ HEALTHSTATE + 0.105* \ LRESLOSS$$
$$(-1.255) \qquad\qquad (2.027)$$

$$+ \ 0.951 \ INFOSCALE + 6.182** \ LRINC800$$
$$(0.963) \qquad\qquad (5.073)$$

$$+ \ 0.597* \ LREQ3\% + 3.253^{(*)} \ DYMEET$$
$$(2.575) \qquad\qquad (1.717)$$

$$+ \ 2.295** \ LLIVTIME + 4.257* \ LNUMCHILD$$
$$(2.689) \qquad\qquad (2.389)$$

$$- \ 9.023* \ FREESCALE - 2.586** \ PROTEST$$
$$(-2.103) \qquad\qquad (-7.239)$$

$$R^2 = 0.554; \ \bar{R}^2 = 0.528; \ d.f. = 189; \ F = 21.295. \qquad (8)$$

Adding the PROTEST variable drastically improves the goodness of fit (R^2 increases by 12 percentage points). The respective coefficient has the expected negative sign and is significant at the 99 percent confidence level. However, the size of most other coefficients decreases and both health variables as well as INFOSCALE even become insignificant (though their t-values indicate that they, nevertheless, add to the explanation of the contributions). On the other hand, LLIVTIME gains in size as well as in statistical significance. This change in the structure of the results is not due to technical rea-

---

32. The three reasons selected by us are the following: "I did not participate because the government should settle affairs like these"; "because the Frenchmen are not entitled to construct an incinerator which affects our well being"; "because giving in by solving the problem at our own expenses would establish a precedent in the future."

sons,[33] but presumably originates in the construction of the protest variable, which is not independent of the endogenous variable (only zero bidders were asked for potential reasons why they did not contribute a [higher] positive amount). Thus, *PROTEST* may predominantly reflect the break between contributors and noncontributors and to a smaller degree genuine protest behavior.[34]

An alternative to protest (a precondition for voice) consists in exit, i.e., in moving to another municipality which remains unaffected by the negative externalities generated by the incinerator. Therefore, in the interview we checked for the willingness to exercise that option. However, we do not obtain any empirical results indicating that the exit option has been seen as a valid possibility. Of course, objections can be raised as to the preselection bias inevitably inherent in our sample, as only those who did not emigrate were questioned. Yet, studying emigration and immigration behavior in Kleinblittersdorf, compared to a large number of other municipalities well before and after the announcement of the plan to erect the incinerator in Grosbliederstroff, yielded no evidence of significant migration activities due to the announced incinerator.[35]

4. Robustness of the Results

In order to estimate a contribution function we have so far been using OLS techniques. Due to the fact that the willingness to contribute is limited to the non-negative range, we proceeded by employing a logarithmic specification. Although the truncation problem is taken care of by applying this specification, a remaining problem consists in the presence of zero contributions, which is "solved" by adding an incremental amount. A more organic way of dealing with the latter problem, which, by the way, allows for checking the robustness of the former results, consists in applying maximum likelihood techniques of estimation. In our context, survival analysis is the most adequate procedure (Hanemann, Loomis, and Kanninen 1990, 11).

In survival models, the dependent variable is restricted to the non-negative range where zero values are explicitly permitted.[36] As a consequence, we are not forced to transform the dependent (contribution) variable,

---

33. The calculated eigenvalues indicate that there is no evidence of multicollinearity; the square root of the ratio of the maximum to the minimum standardized eigenvalue amounts to 11.31 and is, thus, well below the critical threshold of 20 (see Greene 1990, 280).

34. This presumption is nourished by having a look at the interrelation between stated protest behavior of noncontributors and their respective assessment of the odds of success to prevent the incinerator by way of trial. The Spearman rank order correlation (0.199) is far from the expected high confidence level by any conventional measure.

35. This is in line with Hirschman's (1993, 197) idea that there might not only be physical barriers to exit, but also personal countermotives.

36. See Heckmann and Singer 1984, 274.

but can make use of the full information of that variable. We decide to apply an exponential survival model for a number of reasons. First, it allows for multivariate analysis (in analogy with least-squares techniques). Second, the properties of the likelihood function are known to be well behaved (cf. Kiefer 1988, 666). Third, we are therefore in a position to apply indirect (gradient) methods of estimation which, as a by-product, yield estimates of the variance-covariance matrix as well as of the t-values (Gross and Clark 1975, 219). As pointed out by Cox and Oaks (1984, 81), the specification itself does not necessarily play a crucial role once we include covariates in the set up. Therefore, applying the computationally simple exponential model makes good sense, since we express the "hazard rate" as a function of regression factors.[37] The higher the hazard rate, the lower the probability for an individual to exhibit a high willingness to contribute.[38]

Referring to equation (7) in section 3 as the most comprehensive estimation (not burdened with technical problems), the corresponding maximum likelihood estimation equation reads as displayed in table 1.[39]

Since the hazard rate is inversely related to the willingness to contribute, a positive (negative) sign of a coefficient in equation (7) should correspond to a negative (positive) sign in equation (7*). As the results indicate, this indeed is true with the exception of *HEALTHSTATE*, the only variable which is not statistically significant. Analogously to equation (7), the degree of information, both appeal variables, the participation in town meeting, and the variable for free-riding attitude exert a strongly significant impact on an individual's contribution. Moreover, the economic variables in total (especially expected real estate loss, but also expected health damage) play a much stronger role in explaining why people contribute.[40]

5. Conclusions

In this chapter we report on a real world (nearly) Coase-like situation where the citizens of Kleinblittersdorf prevented through voluntary contributions the

---

37. Due to the restriction that the hazard rate has to be nonnegative, an exponential specification seems to be the most plausible choice; see Kalbfleisch and Prentice 1980, 31.

38. Although in the exponential model a continuous distribution of the willingness to contribute is assumed, specifying the setup as a regression estimation model alleviates the problem of having quite a number of zero contributions. Nonparametric survival estimation, which does not rely on the assumption of continuity, is a less attractive alternative because it does not allow for multivariate analysis, but only for the study of the effects of a single treatment.

39. A complete set of estimation results will be supplied by the authors on request.

40. Adding the protest variable to equation (7*) does not change the structure of the estimated coefficients. The log-likelihood decreases (in absolute terms) from 1,333.233 to 866.941 indicating prima facie a better result. However, this seems to be due to the strong correlation between revealed willingness to contribute and the protest variable (the Spearman rank order coefficient is −0.549).

TABLE 1

Log Hazard Rate	Equation (7*)
Constant term	−3.508
HEALTHLOSS	−0.141**
	(−2.742)
HEALTHSTATE	−0.060
	(−1.126)
LRESLOSS	−0.040**
	(−8.696)
INFOSCALE	−0.549**
	(−5.672)
LRINC800	−0.621**
	(−5.428)
LREQ3%	−0.108**
	(−4.795)
DYMEET	−0.852**
	(−4.538)
LLIVTIME	−0.198*
	(−1.789)
LNUMCHILD	−0.466**
	(−2.688)
FREESCALE	2.003**
	(4.899)
Log likelihood	−1,333.233
Wald-statistic	338.179
d.f.	10

construction of a waste incinerator, whose operation would have constituted a (local) public bad. This fact alone casts severe doubt on the absolute validity of the free-rider hypothesis in its strong version.[41] These doubts are substantiated through an econometric analysis of a random sample of 201 citizens. The analysis shows that the individuals' decisions whether and how much to contribute are influenced by economic factors, i.e., in particular by expected health damage and expected real estate value losses, but also by considerations of setup costs in the wave of an opting for exit. However, we should

---

41. The same result is derived by Sandler (in this volume) in his game-theoretic analysis of carbon emissions. Nevertheless, there are some differences between the two chapters. While Sandler focuses on interactions between collectives of nations to provide a public good, the current chapter focuses on interactions within a community to raise contributions for a public good. Whereas the Kleinblittersdorf case study stresses Coase-like cooperative game concerns in terms of empirical testing, Sandler emphasises noncooperative game concerns and their theoretical implications. Besides the free-rider issue which is addressed by both chapters, Sandler studies unilateral actions, neutrality, and alternative strategic assumptions.

stay away from generalizing these results without taking account of the characterization of the particular situation in which the voluntary contribution came about. In this context, it is the influence of institutional factors which should be appreciated first of all. Thus, it is quite obvious from the estimation results that there have been some important coordination efforts which helped to transform the original impersonal large-group setting into a quasi private decision-making framework.[42] Moreover, the option to hold a meeting of all concerned citizens certainly turned out to have been a reinforcing factor. The same is true for embeddedness variables, which may reflect, in part, a process of shaping one's macropreferences to stay and do whatever can be done to accommodate the situation one is locked in at one's place.[43]

Summing up our results, we abstain from the conclusion that free-rider aspects are irrelevant for the provision of public goods. Rather, the message of our paper consists in the idea that in a decentralized setting people are able to handle the free-rider problem on their own. The more confined the (negative/positive) externalities, the more easily the provision of a public good can be left to the citizens, given the framework of participation options on a local level. This not only holds for voluntary contributions to provide a public good, but may also be relevant in the context of public provision financed out of coercive taxes. At least the result of our interviews reveal that not only the majority of citizens/taxpayers, who do not have any real estate, but also a vast majority of real estate owners regarded an increased taxation of real estate as an acceptable solution, preferred to, e.g., lump sum taxation.[44]

It is not implied by our study that coercive solutions to the free-rider problem are not needed at all. Rather, depending on the characteristics of the public good in question, a distinction should be made between public goods that coincide with local jurisdictions and those which exceed this particular framework. Even in the latter case there may be some prospect for successful internalization of externalities. However, in the case of widespread external effects, there is certainly need for a more centralized authority.

---

42. Isaac and Walker (1988) also found (experimental) evidence for the positive influence of communication on the reduction of free-riding behavior.

43. Becker (1992, 340) acknowledges the validity of this argument when stating, "Some of you might be surprised to hear a co-author of the 'de gustibus' point of view, with its emphasis on stable preferences, waxing about the formation of preferences." He calls the phenomenon in question "unthinking attachment."

44. From the characteristics of the German tax system, which is marked by joint taxation on all three government levels, with the consequence that only a small part of total local tax revenue (in 1990 less than 15 percent) stems from own local taxes, we may conclude in general terms that a containment of joint taxation and a strengthening of the tax authority at subfederal levels is feasible.

APPENDIX A: DESCRIPTIVE STATISTICS

Variables	Mean (Proportion)	Median	Standard Deviation	Minimum	Maximum
CON	647.45	50.00	1439.17	0.00	10,895.78
LCON	−8.613	3.912	15.896	−25.328	9.296
LHHINC	8.133	8.161	0.566	6.908	9.328
HEALTHLOSS		2		0	6
HEALTHSTATE		5		0	6
DYCHILD	0.318			0	1
LRESLOSS	0.966	11.290	16.800	−25.328	14.058
LRESTATE	5.961	12.612	14.661	−25.328	14.752
INFOSCALE		4		0	4
LRINC800	0.336	0.000	0.692	0.000	2.644
LREQ3%	0.888	0.000	3.466	0.000	19.905
DYMEET	0.363			0	1
LLIVTIME	3.029	3.332	1.004	0.000	4.317
LNUMCHILD	0.309	0.000	0.465	0.000	1.386
FREESCALE	0.874	0.938	0.185	0.000	1.000
PROTEST		0		0	9

APPENDIX B: CORRELATIONS BETWEEN THE INDEPENDENT VARIABLES

	LHHINC[a]	HEALTHLOSS[b]	HEALTHSTATE[b]	DYCHILD[c]	LRESLOSS[a]	LRESTATE[a]	INFOSCALE[b]
1. LHHINC[a]	1.000						
2. HEALTHLOSS[b]	0.031	1.000					
3. HEALTHSTATE[b]	0.159	0.405	1.000				
4. DYCHILD[c]	0.485	0.145	0.369	1.000			
5. LRESLOSS[a]	0.264	0.168	−0.015	0.728	1.000		
6. LRESTATE[a]	0.193	0.014	−0.070	0.627	0.738	1.000	
7. INFOSCALE[b]	−0.041	0.243	−0.023	0.095	0.252	0.193	1.000
8. LRINC800[a]	0.262	0.015	−0.013	0.379	0.238	0.201	0.166
9. LREQ3%[a]	0.201	0.131	0.022	0.232	0.167	0.122	0.145
10. DYMEET[c]	0.361	0.400	0.248	0.102	0.713	0.627	0.649
11. LLIVTIME[a]	−0.283	0.103	−0.106	0.699	0.295	0.388	0.162
12. LNUMCHILD[a]	0.264	0.068	0.261	0.994	0.150	0.156	0.027
13. FREESCALE[b]	0.178	0.003	0.135	0.553	−0.115	−0.095	−0.074
14. PROTEST[b]	−0.141	−0.125	0.072	0.255	−0.234	−0.193	−0.255

[a]Pearson's coefficients of correlation (if not overridden by contingency or rank order coefficient).
[b]Spearman's rank order coefficients (if not overridden by contingency coefficient).
[c]Corrected Pearson's contingency coefficients.

(continued)

APPENDIX B — *CONTINUED*

	LRINC800[a]	LREQ3%[a]	DYMEET[c]	LLIVTIME[a]	LNUMCHILD[a]	FREESCALE[b]	PROTEST[b]
1. LHHINC[a]							
2. HEALTHLOSS[b]							
3. HEALTHSTATE[b]							
4. DYCHILD[c]							
5. LRESLOSS[a]							
6. LRESTATE[a]							
7. INFOSCALE[b]							
8. LRINC800[a]	1.000						
9. LREQ3%[a]	0.145	1.000					
10. DYMEET[c]	0.466	0.351	1.000				
11. LLIVTIME[a]	0.068	−0.029	0.750	1.000			
12. LNUMCHILD[a]	0.163	−0.012	0.092	−0.109	1.000		
13. FREESCALE[b]	−0.161	−0.023	0.524	−0.168	0.063	1.000	
14. PROTEST[b]	−0.298	−0.155	0.424	0.090	−0.105	0.164	1.000

REFERENCES

Andreoni, James. 1988. Why Free Ride? Strategies and Learning in Public Goods Experiments. *Journal of Public Economics* 37:291–304.

Axelrod, Robert. 1984. *The Evolution of Cooperation.* New York: Basic Books.

Bagnoli, Mark, and Michael McKee. 1991. Voluntary Contribution Games: Efficient Private Provision of Public Goods. *Economic Inquiry* 29:351–66.

Barro, Robert J. 1974. Are Government Bonds Net Wealth? *Journal of Political Economy* 82:1095–1117.

Becker, Gary S. 1974. A Theory of Social Interactions. *Journal of Political Economy* 82:1063–93.

———. 1992. Habits, Addictions and Traditions. *Kyklos* 45:327–46.

Bohm, Peter. 1972. Estimating the Demand for Public Goods: An Experiment. *European Economic Review* 3:111–30.

Bolnick, Bruce R. 1976. Collective Goods Provision through Community Development. *Economic Development and Cultural Change* 25:137–50.

Brubaker, Earl R. 1982. Sixty-Eight Percent Free Revelation and Thirty-Two Percent Free Ride? Demand Disclosures under Varying Conditions of Exclusion. In Vernon L. Smith, ed., *Research in Experimental Economics* 2:151–66. Greenwich, Conn.: JAI Press.

———. 1984. Demand Disclosures and Conditions on Exclusion: An Experiment. *Economic Journal* 94:536–53.

Cox, David R., and D. Oaks. 1984. *Analysis of Survival Data.* London and New York: Chapman and Hall.

Gaitsgory, Vladimir, and Shmuel Nitzan. 1993. A "Folk-Theorem" in a Dynamic Game of Private Provision of Public Goods. Manuscript, Bar Ilan University, Ramat Gan.

Greene, William H. 1990. *Econometric Analysis*. New York: Macmillan.

Gross, Alan J., and Virginia Clark. 1975. *Survival Distributions*. New York: John Wiley and Sons.

Hanemann, W. Michael; John Loomis; and Barbara Kanninen. 1990. Statistical Efficiency of Double-Bounded Dichotomous Choice Contingent Valuation. Working paper, University of California, Berkeley.

Hardin, Garrett. 1968. The Tragedy of the Commons. *Science* 162:1243–48.

Heckman, James, and Burton Singer. 1984. A Method for Minimizing the Impact of Distributional Assumptions in Econometric Models for Duration Data. *Econometrica* 52:271–320.

Hirschman, Albert O. 1982. *Shifting Involvements: Private Interest and Public Action*. Princeton, N.J.: Princeton University Press.

———. 1993. Exit, Voice and the Fate of the German Democratic Republic: An Essay in Conceptual History. *World Politics* 45:173–202.

Isaac, Mark R.; Kenneth F. McCue; and Charles R. Plott. 1985. Public Goods Provision in an Experimental Environment. *Journal of Public Economics* 26:51–74.

Isaac, Mark R., and James M. Walker. 1988. Communication and Free Riding Behaviour: The Voluntary Contribution Mechanism. *Economic Inquiry* 26:585–608.

Kahneman, Daniel, and Amos Tversky. 1979. Prospect Theory: An Analysis of Decision under Risk. *Econometrica* 47:263–91.

Kalbfleisch, John D., and Ross L. Prentice. 1980. *The Statistical Analysis of Failure Time Data*. New York: John Wiley and Sons.

Kiefer, Nicholas M. 1988. Economic Duration Data and Hazard Functions. *Journal of Economic Literature* 26:646–79.

Kikuchi, Masao; Geronimo Dozina, Jr.; and Yujiro Hayami. 1978. Economics of Community Work Programs: A Communal Irrigation Project in the Philippines. *Economic Development and Cultural Change* 26:211–25.

Klein, Daniel B. 1990. The Voluntary Provision of Public Goods? The Turnpike Companies of Early America. *Economic Inquiry* 28:788–812.

Margolis, Howard. 1982. *Selfishness, Altruism and Rationality: A Theory of Social Choice*. Cambridge: Cambridge University Press.

Mestelman, Stuart, and David Feeny. 1988. Does Ideology Matter? Anecdotal Experimental Evidence on the Voluntary Provision of Public Goods. *Public Choice* 57:281–86.

Musgrave, Richard A., and Peggy B. Musgrave. 1984. *Public Finance in Theory and Practice*. New York: McGraw-Hill.

Ng, Yew-Kwang. 1988. Economic Efficiency versus Egalitarian Rights. *Kyklos* 41:215–37.

Okun, Arthur M. 1975. *Equality and Efficiency, the Big Tradeoff*. Washington, D.C.: Brookings Institution.

Olson, Mancur. 1965. *The Logic of Collective Action: Public Goods and the Theory of Groups*. Cambridge, Mass.: Harvard University Press.

Ostrom, Elinor. 1990. *Governing the Commons: The Evolution of Institutions of Collective Action*. Cambridge: Cambridge University Press.

Palfrey, Thomas R., and Howard Rosenthal. 1988. Private Incentives in Social Dilemmas: The Effects of Incomplete Information and Altruism. *Journal of Public Economics* 35:309–32.

Pommerehne, Werner W.; Lars P. Feld; and Albert Hart. 1994. Voluntary Provision of a Public Good: Results from a Real World Experiment. *Kyklos* 47:505–18.

Samuelson, Paul A. 1954. Pure Theory of Public Expenditure. *Review of Economics and Statistics* 36:387–89.

Sandler, Todd. 1995. A Game-Theoretic Analysis of Carbon Emissions. In this volume.

Schiff, Jerald. 1990. *Charitable Giving and Government Policy: An Economic Analysis*. New York: Greenwood Press.

Schneider, Friedrich, and Werner W. Pommerehne. 1981. Free Riding and Collective Action: An Experiment in Public Microeconomics. *Quarterly Journal of Economics* 96:689–704.

Smith, Vernon L. 1979. An Experimental Comparison of Three Public Good Decision Mechanisms. *Scandinavian Journal of Economics* 81:198–215.

Stigler, George J. 1961. The Economics of Information. *Journal of Political Economy* 69:213–25.

Van de Kragt, Alphons J. C.; Robyn M. Dawes; and John M. Orbell. 1988. Are People Who Cooperate "Rational Altruists"? *Public Choice* 56:233–47.

Weisbrod, Burton A. 1975. Toward a Theory of the Voluntary Non-Profit Sector in a Three-Sector Economy. In Edmunds S. Phelps, ed., *Altruism, Morality and Economic Theory*, 171–95. New York: Russell Sage.

Weisbrod, Burton A., and Nestor D. Dominguez. 1986. Demand for Collective Goods in Private Non-Profit Markets: Can Fundraising Expenditures Help Overcome Free Riding Behaviour? *Journal of Public Economics* 30:83–95.

Wilson, Loretta S. 1992. The Harambee Movement and Efficient Public Good Provision in Kenya. *Journal of Public Economics* 48:1–19.

CHAPTER 11

A Game-Theoretic Analysis of Carbon Emissions

Todd Sandler

Introduction

Many of the world's most precious assets are owned in common by all nations and, as such, are borderless. Examples include the stratospheric ozone shield, the oceans (beyond 200 miles from coastlines), and the troposphere. Other assets, while providing benefits to current and future generations worldwide, are located within political borders. The moist tropical forests and barrier reefs with their vast biodiversity are apt examples. The first class of assets is associated with global commons problems, for which open access often re-sults in a tragedy of the commons as the resource or asset is overexploited by users who do not consider the costs that their actions impose on others. The second class of assets may also be inefficiently exploited if the nation hosting the resource does not account for transnational public benefits derived from the resource. If, for example, the host nation includes only private and public benefits received within its own territory when making allocative decisions, inefficiency will result at the global level.

Global warming involves the atmospheric accumulation of carbon and includes more than the problem of burning fossil fuels, which accounts for approximately half of anthropogenic atmospheric heating.[1] At least, three other global commons problems are related; each possesses its own distribu-tion of costs and benefits (Sandler 1992). First, the preservation of the ozone shield comes into play, since chlorofluorocarbons (CFCs), which deplete ultraviolet-shielding ozone, are greenhouse gases (GHGs). In 1987, the major

This chapter was prepared under a cooperative agreement between the Institute for Policy Reform (IPR) and the Agency for International Development, cooperative agreement no. PDC-0095-A-00-1126-00. Additional funding was provided by the National Science Foundation grant no. SBR-9222953. The author has benefited from the comments of two anonymous ref-erees. The views expressed are solely those of the author, who is a senior fellow of IPR and a distinguished professor of economics at Iowa State University.

1. On global warming, see Morgenstern 1991; Morrisette and Plantinga 1991; Nordhaus 1991; and White 1990.

emitters of CFCs (in descending order) were the United States, the USSR, Japan, West Germany, Italy, the United Kingdom, and France[2]; they accounted for two-thirds of all CFCs emissions. Second, population growth leads to food demands, fuel needs, and wastes that release methane. The six largest emitters (i.e., the United States, India, China, the USSR, Canada, and Brazil) contributed over half of all methane emissions. Third, tropical forests are disappearing at a rate of 20.4 million hectares per year (World Resources Institute 1990, chap. 7). The world's forests are an important sink for carbon, storing approximately 450 billion metric tons. In fact, tropical deforestation may now account for upwards of one-third of the accumulation of atmospheric carbon. Unlike the methane and CFCs problem, deforestation is almost entirely taking place in less developed countries (LDCs). When the various sources of GHGs are taken into account, GHG emitters include both developed and less developed countries, and each set of nations must be considered.

The atmospheric heating problem provides challenges to the economic modeler. First, the interdependency among the contributing commons problems must be taken into account. Second, the strategic interactions among nations worldwide must be included. Standard prescriptions, such as Pigovian taxes or Coase bargaining, for externalities will not work if transnational strategic interactions are ignored. Hence, a carbon tax on a barrel of oil imposed at the national level may have little ameliorating influence when other nations react in an optimizing fashion. Third, cooperative supranational pacts and agreements that involve a *subset* of nations must be judged in light of the responses from nations outside the pact.[3] Finally, both nation-specific benefits (outputs) and transnational public benefits (outputs) must be included when assessing the international distribution of costs and benefits that arise from activities leading to global warming. That is, joint products are often involved and must be included in the model.

The purpose of this chapter is to devise a simple game theory representation of the carbon emissions problem that includes the interdependencies among contributing commons problems, interactions among nations, and the presence of joint products. Although the focus is on global warming, the framework is designed to be tractable yet sufficiently flexible to analyze a host of different transnational common property scenarios. A second purpose is to employ the framework to investigate the outcome of unilateral action and/or cooperative pacts among a subset of nations. Unilateral action occurs when an

---

2. The fact from this paragraph are derived from table 24.2 in World Resources Institute 1990, p. 348–49.

3. On this issue, see the important studies by Barrett (1991), Carraro and Siniscalco (1992), Hoel (1991), and Mäler (1991).

agent responds to a change in its parameters (e.g., prices) or else acquires new concerns, such as altruistic interests.

Initially, two sets of coalitions of nations are assumed, so that I can first identify the strategic interaction at the supranational level between coalitions. The models consist of simultaneous-play noncooperative games between two or more agents that seek either to provide a public good through pollution reduction, or else to pursue an agent-specific (private) benefit from an economic activity. In the latter case, the activity also gives off a public bad (i.e., atmospheric carbon) as a jointly produced output. After presenting the case of unified coalitions, I consider the implications if coalitions are not unified. The analysis can identify a number of pitfalls of policy and can also be generalized in future work to account for intertemporal aspects of carbon accumulation.[4]

Pure Public Good Model

To highlight the distinguishing features of joint products in the case of global warming, I first consider the pure public good model, which can correspond to a scenario where nations remove transfrontier pollution from a commons or shared ecosystem. In the case of global warming, atmospheric carbon removal can be achieved through large-scale tropical reforestation projects, since the fast-growing trees can sequester carbon. Moreover, the iron content of oceans can be raised to augment sequestered carbon levels. The pure public good model also applies to scenarios in which nations curb their carbon emissions either by installing equipment in power plants to trap carbon or else by increasing energy efficiency and conservation.

The model includes two coalitions of countries: the developed countries (D) and the less developed countries (L). This partition of nations is chosen to highlight, at a later stage, the effects of income redistribution or foreign aid from the developed to the less developed nations. Each coalition is initially treated as a unified actor, whose utility is derived from a private numéraire good, y, and a pure public good, Z. The latter denotes pollution removal and equals aggregate pollution removal:

$$Z = z^D + z^L, \tag{1}$$

where z^i, $i = D, L$, indicates the respective coalition's pollution reduction. Each coalition's preferences are represented by a quasiconcave, strictly increasing utility function:

$$U^i = U^i(y^i, Z), \qquad i = D, L. \tag{2}$$

---

4. Intertemporal models include Nordhaus 1991 and Ko, Lapan, and Sandler 1992.

The coalition's budget or resource constraint is[5]

$$I^i = y^i + pz^i, \qquad i = D, L, \tag{3}$$

where I^i is coalition i's income endowment, p is the per-unit price of pollution removal, and unity is the per-unit price of the numéraire. Each coalition chooses y^i and z^i to

$$\text{maximize } \{U^i(y^i, z^i + z^j) \mid I^i = y^i + pz^i\}, \tag{4}$$

where z^j is treated as j's best response and is fixed (exogenous) to coalition i.

For the developed nations coalition, the first-order conditions (FOCs) associated with (4) require the following equality:

$$MRS^D_{Zy} = p, \tag{5}$$

in which MRS^D_{Zy} is coalition D's marginal rate of substitution between pollution removal and the private good, and, as such, denotes the marginal benefit from the public good. The MRS is evaluated at the best-response or optimizing level of z^L. An analogous FOC characterizes the less developed nations coalition. A Nash equilibrium is attained for the two coalitions when a vector of pollution removal amounts, (z^D_N, z^L_N), is achieved that simultaneously satisfies the FOCs. Subscript N denotes a Nash equilibrium.

The resulting Nash equilibrium is Pareto suboptimal since benefits conferred on the other coalition by a coalition's action are not taken into account. Pareto optimality requires each coalition to remove pollution until

$$MRS^D_{Zy} + MRS^L_{Zy} = p \tag{6}$$

is satisfied.[6]

To illustrate the suboptimality of the Nash equilibrium and other results, I employ a graphical apparatus introduced by Cornes and Sandler (1984, 1986). From (4), budget-constrained utility for coalition D can be written as

$$U^D = U^D(I^D - pz^D, z^D + z^L). \tag{7}$$

----

5. A nonlinear convex constraint could also be used to allow marginal costs to vary with pollution abatement.

6. The underlying problem is

$$\underset{y^D, y^L, z^D, z^L}{\text{maximize}} \ \{U^D(y^D, Z) \mid U^L(y^L, Z) \geq \bar{u}^L, \ Y + pZ = I\},$$

where $\bar{u}^L$ is a given level of utility for coalition L, $Y = y^D + y^L$, and $I = I^D + I^L$.

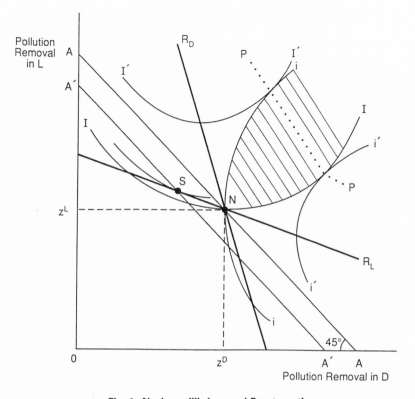

Fig. 1. Nash equilibrium and Pareto path

Given income and relative prices, equation (7) indicates that utility can be depicted in terms of pollution removal levels in the two coalitions. The set of points weakly preferred by coalition D to any reference allocation must be a convex set, since it is itself the intersection of two convex sets: points on or below the budget constraint, and points on or above an indifference curve.

In figure 1, curves II and I'I' depict two budget-constrained iso-utility curves for coalition D, where z^L is measured on the vertical axis and z^D on the horizontal axis. Iso-utility curve I'I' denotes a higher level of utility than II, since for a given z^D, the amount of coalition L's pollution removal is greater. By the implicit function rule applied to (7), the slope of the iso-utility curve is

$$dz^L/dz^D = -1 + (p/MRS^D_{Zy}). \tag{8}$$

When this slope is zero, the FOC in (5) is satisfied for a given level of pollution removal in coalition L. A Nash reaction path is derived by joining the zero-sloped points of the iso-utility curves for different levels of z^L. Path

R_D denotes the path for coalition D. If all goods are normal with positive income elasticities less than one, which we assume henceforth, then the slope of the reaction path must be negative and greater than one in absolute value.[7]

An analogous procedure can be used to display coalition L's budget-constrained iso-utility curves in figure 1. Two of these curves—ii and i'i'— are depicted. These curves are U-shaped oriented to the vertical axis, and have slopes equal to the reciprocal of $[-1 + (p/MRS^L_{Zy})]$. The Nash reaction path for coalition L is R_L in figure 1. When all goods are normal, the slope of R_L must be negative and less than one in absolute value. For normal goods, the unique Nash equilibrium is at point N where both coalitions' Nash FOCs are mutually satisfied; hence, neither coalition has an incentive to alter its provision of the public good. This Nash equilibrium is Pareto-inferior to points in the cross-hatched region, formed by the iso-utility curves through N, since both coalitions' utility would increase in this region. The Pareto-optimal path, PP, corresponds to the tangencies between the coalitions' iso-utility curves, where $\Sigma MRS^i_{Zy} = p$ is satisfied. To show the total amount of pollution removed at the Nash equilibrium, draw line AA with slope -1 through point N, so that this line makes a 45 degree angle with the two axes. Distance OA along either axis then measures the total pollution removed by the two coalitions.

If one coalition were to move first as a Stackelberg leader, and the other were to follow, then the leader would gain an advantage, while the total amount of pollution removed would decline. Suppose that coalition D is the leader and coalition L is the follower. Coalition D maximizes its utility subject to its budget constraint and to the reaction path of the follower. In figure 1, this equilibrium corresponds to point S, where coalition D's iso-utility curve is tangent to R_L. By inspection, we see that the leader's (follower's) utility is greater (smaller) at point S than at point N. There is an advantage to assuming a Stackelberg leadership role. The 45 degree line A'A' indicates that total pollution removal has fallen.

Is it in a coalition's (or nation's) interest to set a "good example" by *unilaterally* instituting pollution abatement or removal actions beyond those associated with an initial Nash equilibrium when a global pollution problem is confronted? If unilateral action is interpreted in terms of the datum of the original Nash problem, then the analysis is somewhat trivial. This follows

---

7. This follows from the comparative static derivative,

$$\partial z^D_N / \partial z^L = p \partial z^D_N / \partial I^D - 1,$$

which indicates the effect of a change in z^L on the Nash equilibrium level of z^D. The right-hand expressions are an income effect and a substitution effect, respectively. The latter is equal to -1, since pollution removal is a perfect substitute regardless of which coalition performs the cleanup. Reaction paths do not have to be linear, but have been drawn that way for convenience.

because neither coalition has, by definition, an incentive at a Nash equilibrium to alter its removal activity. Hence, any type of unilateral action for the original game structure would be welfare reducing for the coalition. To present a more interesting case of unilateral action, I interpret this action as arising from a change in a parameter confronting one coalition, so that a *new* Nash equilibrium applies. Alternatively, unilateral action could involve a situation in which one of the coalitions alters its preferences so as to include altruistic interests. Both of these cases have essentially identical outcomes.

Suppose first that coalition D attempts to set a "good example" by subsidizing pollution removal with a per-unit subsidy, t, financed by a lump-sum tax of T. This corresponds to coalition D choosing z^D and y^D to

$$\text{maximize } \{U^D(y^D, z^D + z^L) \mid I^D - T = y^D + (p - t)z^D\}, \tag{9}$$

where z^L is coalition L's best response. The FOC is now

$$MRS^D_{Zy} = p - t, \tag{10}$$

which implies (owing to quasi-concave utility) a greater amount of pollution removal for each level of z^L. This result is illustrated in figure 2, where point N is the Nash equilibrium, prior to the subsidization scheme. Once in place, the unilateral subsidization scheme shifts coalition D's reaction path rightward to R'_D. The latter curve connects positively sloped points on the iso-utility curve where (10) is satisfied.[8]

At the new Nash equilibrium M, total pollution control has increased from OA to OB; however, the efforts of coalition L have diminished relative to the original Nash equilibrium. Coalition L has taken advantage of coalition D's augmented efforts, and this results in coalition D being worse off at M than at N since I'I' is a lower level of welfare than II. Setting a good example does not pay in the case of pure public good.[9]

Next suppose that the price of pollution removal drops to $\bar{p}$, owing to a technological breakthrough that affects just coalition D. Unlike the first case, the price fall is not engineered through a tax policy. For each level of z^L, coalition D would now satisfy $MRS^D_{Zy} = \bar{p}$. Since $\bar{p} < p$, this situation again gives a reaction path such as R'_D to the right of R_D in figure 2. Coalition D

---

8. The same iso-utility map can be used, since I^D, p, and tastes are unchanged. Furthermore, $dT = z^D dt$ so that the subsidy is fully financed by the lump-sum tax, thus eliminating the income effect.

9. Hoel (1991) uses a similar, but different, model to demonstrate that unilateral action may reduce the well-being of the activist nation. Hoel's interesting analysis also examines cooperative behavior.

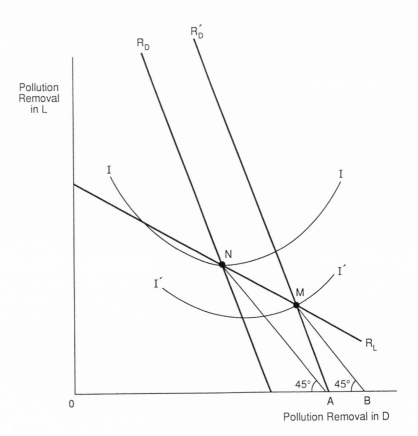

Fig. 2. Unilateral action

augments its pollution reduction, coalition L reduces its effort, and the overall effort increases. Ironically, the technological breakthrough places coalition D onto a lower indifference curve because of coalition L's strategic reaction.

Because price has now changed without a financing offset, a new set of iso-utility curves would apply, so that reaction path R_D' connects the zero-sloped points on this new iso-utility map (not shown). The influence of the price drop on D's utility is not known without more structure being placed on the problem. If the prices faced by coalition L do not change, as assumed, then L would experience an increase in utility at point M as it relies more on D's pollution abatement. If coalition D's utility were to drop owing to the technological breakthrough, then it would be in its interest to keep it secret.

Alternatively, suppose that coalition D acts unilaterally because it begins to take an interest in coalition L's well-being so that coalition D now maximizes

$$U^D(y^D, Z) + \psi U^L(y^L, Z),$$ (11)

where $\psi \in (0, 1)$. Along coalition D's new reaction path, the following equality is satisfied:

$$MRS^D_{Zy} + \psi[(\partial U^L/\partial Z)/(\partial U^D/\partial y^D)] = p.$$ (12)

Given quasiconcave utility functions, (12) implies a greater z^D for each z^L and a consequential rightward shift of the reaction path R_D. The effects on pollution removal is the same as the other unilateral action cases. Nevertheless, we cannot necessarily conclude that D is worse off at the new Nash equilibrium, since its tastes have changed as compared to the original equilibrium. Two different iso-utility maps would apply to a figure like figure 2.

Until now, the constituent nations of each coalition are assumed to act cooperatively vis-à-vis one another. This may be a reasonable assumption in the case of a coalition, such as the European Community (EC), if the nations unite politically and a unanimous decision rule is imposed. If, however, nations of the coalition do not cooperate, then paradoxically the coalition *may* actually improve its well-being. Assume, for simplicity, that coalition D is composed of just two countries that behave in a noncooperative fashion, while coalition L is assumed unified. For coalition D, *each* nation must choose z^D_i and y^D_i, $i = 1, 2$, to

$$\text{maximize } \{U^D_i(y^D_i, z^D_i + z^D_j + z^L) \mid I^D_i = y^D_i + pz^D_i\},$$ (13)

where both z^D_j and z^L are exogenous, best-response levels. Optimality now requires each nation of coalition D to satisfy

$$MRS^{Di}_{Zy} = p,$$ (14)

in which the i superscript refers to the nation. In the earlier problem, each coalition member satisfies

$$MRS^D_{Zy} = MRS^{D1}_{Zy} + MRS^{D2}_{Zy} = p.$$ (15)

The important implication of noncooperation within the coalition can be displayed by reinterpreting figure 2. Suppose that each nation in coalition

D is identical, with the same endowments and tastes. Further suppose symmetric equilibrium for coalition members. First, let the indifference curves for coalition D denote the within-coalition *cooperative* case with slope equal to

$$dz^L/dz^D = -1 + \left(p \Big/ \sum MRS^{Di}_{Zy} \right),\qquad(16)$$

where $z^D = z^D_1 + z^D_2$. In figure 2, path R_D then connects points of zero slope where (15) is satisfied. For the noncooperative case, equation (14) is, however, fulfilled for each coalition member, so that the equilibrium path must then correspond to points on each iso-utility curve *with negative* slope to the left of the zero-sloped points. This follows because MRS^{Di}_{Zy} is less than the sum in (15). In consequence, the noncooperative path, R''_D, not shown, would lie to the left of R_D and may imply a greater utility level for coalition D at the intersection of R''_D and R_L. Coalition D's welfare improves when this new equilibrium is between point N and the intersection of R_L and curve II. Of course, overall pollution removal falls. This improvement in coalition D's welfare is more likely when R_L is steep and the leftward shift of R_D is not too great. A steeper R_L means that coalition L will increase its pollution removal more fully in reaction to coalition D's reduced effort. A smaller leftward shift of R_D implies less free riding within coalition D. If this free riding becomes sufficiently great, then coalition D may lose welfare despite coalition L's augmented efforts. Thus, cooperation at a subglobal level may be harmful to participants when global public goods are considered (see Bruce 1990 on arms expenditures). Within-coalition cooperation may be desirable only when the major polluters are included, since strategic maneuvering outside of the coalition may then have little impact.

But why would a coalition form if the participants are worse off? Mäler (1991, p. 167) indicated that such a coalition may gain on other fronts not encompassed by the net benefits under study. He gave the example of the Columbia River Treaty involving the United States and Canada, for which the United States appeared to lose. According to Mäler, the United States could gain from longer-range benefits derived from the economic development in Canada. A similar case involved a U.S.-Mexican agreement on the desalinization of the Colorado River. Public choice aspects may also come into play when coalitions form, but some participants suffer. If the officeholder signs an environmental agreement to win political or lobbying support for reelection from a subset of voters, the net benefits to the entire constituency may be negative. Finally, coalitions may give perverse results because rationality in the real world may fall short of the standard required by game theory.

Another interesting case concerns a coalition of developed countries financing pollution reduction in the LDCs. The so-called neutrality theo-

TABLE 1. Summary of Results for Pure Public Good Scenarios

Cases	Pollution Removal in D	Welfare in D	Pollution Removal in L	Welfare in L	Total Pollution Removal
Coalition D is Stackelberg leader	decrease	increase	increase	decrease	decrease
Unilateral action by coalition D	increase	decrease[a]	decrease	increase	increase
Noncooperative behavior within coalition D	decrease	ambiguous	increase	decrease	decrease
Income redistribution[b] from coalition D to coalition L	decrease	no change	increase	no change	no change

[a]Unilateral action involving a change in coalition D's tastes or a technology-induced drop in the price of z^D would have an ambiguous effect on D's welfare.

[b]An income redistribution that does not change the set of contributors.

rem[10] may nullify the desired influence of pollution-reducing aid flowing from the developed countries to the less developed countries when a *global* pollution cleanup is involved. The neutrality theorem indicates that the Nash-equilibrium provision level for a *pure* public good is invariant to income redistributions among an unchanged set of providers. This follows because the aid donor's cutback on pollution removal matches the aid recipient's increased removal, even when the economic agents are not identical, so that strategic interactions may undo well-intentioned policies. If, *however,* the redistribution of income is sufficiently large to alter the set of public good providers, or else flows to countries with no previous pollution-reducing activities, then neutrality does not apply and policy can engineer changes in the overall level of pollution removal (Bergstrom, Blume, and Varian 1986). Other influences, including different comparative advantages in removing pollution among providers, may also nullify neutrality. Aid should then go to those nations with the lowest marginal costs pollution removal.

Table 1 summarizes the primary findings in this section. For each case, changes in pollution removal and welfare levels are characterized.

Joint Product Models of Carbon Accumulation

To capture the essential elements of carbon accumulation in a tractable model,[11] I allow a coalition-specific activity, q^i, to yield a coalition-specific

---

10. The neutrality theorem is discussed in Warr 1983; Cornes and Sandler 1984, 1986; and Sandler 1992 (pp. 77–79).

11. Additional joint products of a public or private nature could be included to account for such things as biodiversity or existence value.

private good, x^i, and atmospheric carbon, z^i. The latter gives rise to global warming, C, which is a global public bad. As before, two coalitions are assumed: developed and less developed nations.

Fixed proportions are assumed to characterize the production of the joint products. For the private (coalition-specific) output, the joint product relationship is

$$x^i = \alpha^i q^i, \qquad i = D, L, \tag{17}$$

where α^i's are positive, and x^D and x^L can denote completely different outputs. Similarly, activity q can differ between the coalitions; hence, q^L can refer to deforestation, while q^D can refer to the burning of fossil fuels or the emission of CFCs. Carbon results from activity q^i as follows:

$$z^i = \beta^i q^i, \qquad i = D, L, \tag{18}$$

where β^i's are positive. Global heating depends on carbon emissions, so that

$$C = C(\beta^D q^D + \beta^L q^L), \tag{19}$$

which is, henceforth, written as the identity map,

$$C = \beta^D q^D + \beta^L q^L \equiv Q, \tag{20}$$

where the right-hand side denotes the overall increase in the carbon index.

The coalition's strictly quasiconcave utility function is

$$U^i = U^i(y^i, \alpha^i q^i, Q), \qquad i = D, L, \tag{21}$$

in which $\partial U^i / \partial y^i > 0$, $\partial U^i / \partial (\alpha^i q^i) > 0$, and $\partial U^i / \partial Q < 0$. The coalition's budget is divided between the private numéraire and the joint-product activity,

$$I^i = y^i + r^i q^i, \qquad i = D, L, \tag{22}$$

where r^i is the per-unit price of the activity.

Each coalition chooses y^i and q^i to maximize utility in (21) subject to its budget constraint and the best-response level of q in the other coalition. For a Nash equilibrium, the FOCs can be written as

$$\alpha^i MRS^i_{xy} + \beta^i MRS^i_{Qy} = r^i, \qquad i = D, L, \tag{23}$$

in which the weighted *MRS*'s are set equal to the relative price of the activity. Since MRS_{Qy}^i is negative, the activity will not take place unless the first term on the left-hand side of (23) exceeds the second in absolute value. A Nash equilibrium corresponds to a vector (q_N^D, q_N^L) that simultaneously satisfies (23) for both coalitions. The Nash equilibrium does not correspond to a Pareto optimum, which for, say, coalition D requires that

$$\alpha^D MRS_{xy}^D + \beta^D MRS_{Qy}^D + \beta^L MRS_{Qy}^L = r^D \qquad (24)$$

for $q^D > 0$. In contrast to the public good case, too much of the activity occurs in the Nash equilibrium as compared with the Pareto optimum, since the negative externality imposed on others is ignored. Also corner solutions are a greater consideration owing to this externality.

The Cornes-Sandler diagrammatic apparatus is now modified to display the case of joint products. The budget-constrained iso-utility function for coalition D is

$$U^D = U^D(I^D - r^D q^D, \ \alpha^D q^D, \ \beta^D q^D + \beta^L q^L) \qquad (25)$$

which can be displayed in (q^D, q^L) space for fixed production coefficients, income, and prices. Since the utility function in (21) is strictly quasiconcave, and since activities and final outputs are related in a linear fashion, budget-constrained utility in (25) must also be quasiconcave, so that the set of points (q^D, q^L) weakly preferred to any given allocation is convex. An increase in q^L, for a given level of q^D, implies a lower utility level as q^L gives forth more global carbon but no benefits for coalition D. In figure 3, two of coalition D's iso-utility curves are displayed, where the cross-hatched region indicates preferred points. The slope of the iso-utility curve is

$$\frac{dq^L}{dq^D} = \frac{r^D - \alpha^D MRS_{xy}^D - \beta^D MRS_{Qy}^D}{\beta^L MRS_{Qy}^D}. \qquad (26)$$

When this slope is zero, the FOCs for a Nash equilibrium are satisfied at a given level of q^L. The Nash reaction path connects the points of zero slope. In figure 3, the Nash reaction path is drawn negatively sloped; but this is only one possibility. For coalition L, the iso-utility curves are hill-shaped contours oriented to the vertical axis.

To derive a general expression for the slope of the reaction path, I follow a procedure presented in Cornes and Sandler (1986, pp. 118–20). Restricted cost functions,

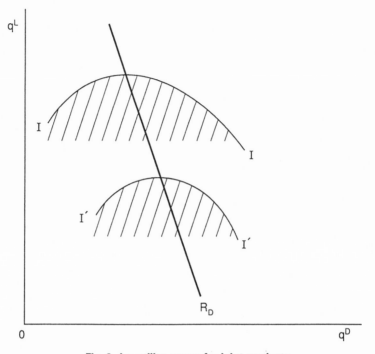

Fig. 3. Iso-utility curves for joint products

$$y^D(x^D, Q, u^D) \equiv \underset{y^D}{\text{minimize}} \ \{y^D \mid U^D(y^D, x^D, Q) \geq u^D\}, \qquad (27)$$

are defined in terms of the Nash equilibrating values of x^D, Q, and u^D. The partial derivatives of $y^D(\cdot)$ with respect to x^D and Q yield compensated inverse demand functions that express the coalition's MRS's as

$$-\partial y^D/\partial x^D = MRS^D_{xy} \equiv \pi^D_x(x^D, Q, u^D), \qquad (28)$$

$$-\partial y^D/\partial Q = MRS^D_{Qy} \equiv \pi^D_Q(x^D, Q, u^D). \qquad (29)$$

These relationships can be substituted into (23) for the MRS expressions. By taking a total differential and rearranging the resulting equation, I get the slope of the reaction path:[12]

---

12. See Cornes and Sandler 1994 for a different decomposition of the slope of the reaction path and an analysis of other comparative static changes.

$$\frac{dq^D}{dq^L} = \left[\frac{\beta^L (\alpha^D \pi^D_{xQ} + \beta^D \pi^D_{QQ})}{\Omega} \right] + \left[\frac{\alpha^D \pi^D_{xu} + \beta^D \pi^D_{Qu}}{\Omega} \right] \frac{du^D}{dq^L}, \quad (30)$$

where $\pi^D_{xQ} = \partial \pi^D_x / \partial Q$, $\pi^D_{QQ} = \partial \pi^D_Q / \partial Q$, etc. In (30), the denominator of the bracketed term is

$$\Omega = -(\alpha^D \ \beta^D) \begin{bmatrix} \pi^D_{xx} & \pi^D_{xQ} \\ \pi^D_{Qx} & \pi^D_{QQ} \end{bmatrix} \begin{pmatrix} \alpha^D \\ \beta^D \end{pmatrix} \geq 0. \quad (31)$$

Its sign follows from the strict quasiconcavity of utility. Henceforth, I assume that Ω and its subcomponents are nonzero.

The first right-hand expression in (30) is an income-compensated substitution effect, while the second expression is an income effect. The sign of the substitution effect is ambiguous, because π^D_{xQ} may be of either sign, depending on the consumption relationship of the jointly produced goods. If, for example, x^D and Q are Hicksian q-substitutes or independent outputs ($\pi^D_{xQ} < 0$ or $= 0$), then the substitution effect must be negative, since π^D_{QQ} is negative by strict quasiconcavity. For Hicksian q-complements, the substitution effect is positive when $\alpha^D \pi^D_{xQ} > |\beta^D \pi^D_{QQ}|$.

The income effect is also ambiguous since both a good and a bad is derived from q^D. The multiplicative factor du^D / dq^L is negative, but π^D_{xu} is positive while π^D_{Qu} is negative. A negative income effect results when $\alpha^D \pi^D_{xu} > |\beta^D \pi^D_{Qu}|$, so that the willingness to pay for x as real income increases is greater than the willingness to pay to avoid more Q or carbon. If, however, increased wealth is associated with a greater income elasticity for the environment, then the income effect may be positive. Apparently, the slope of the reaction path depends on a host of factors and may be negative or positive. Nevertheless, some important cases can be highlighted. In the absence of income effects, substitute joint products are unequivalently associated with negatively sloped reaction paths, while complement joint products are a necessary, but not sufficient, condition for positively sloped reaction paths. If joint products are substitutes or independent of one another, and environmental concern has a small income effect, then negatively sloped reaction paths are expected. If, however, environmental concern increases with income and the public bad is viewed very unfavorably, then the income effect may be positive and dominate the substitution effect. In such a case, positively sloped reaction paths would apply by (30).

Case 1: Negatively Sloped Reaction Paths

In figure 4, the joint output producing activities q^D and q^L are measured on the horizontal and vertical axes, respectively. The initial Nash reaction paths are

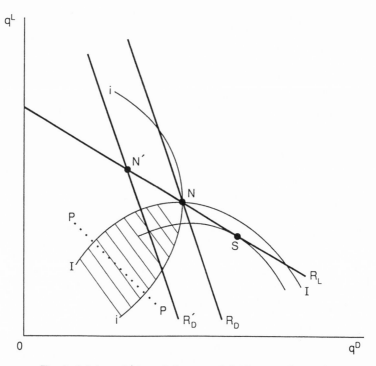

Fig. 4. Joint products and downward sloping reaction paths

R_D and R_L, and the Nash equilibrium is N, which implies a Pareto-inferior position. The Pareto path, PP, connects the tangencies of the iso-utility curves. If unilateral action[13] on behalf of, say, coalition D is performed by instituting a per-unit tax on activity q^D with the money returned through a lump-sum subsidy, then D's reaction path shifts leftward to R_D' and the new equilibrium is N'. Coalition D becomes worse off, while coalition L becomes better off. Given the shapes of the reaction paths, coalition L increases its q^L activity, and thus its output of carbon, in response to coalition D's unilateral decrease in q^D. A 45 degree line through N' and N (not shown) would indicate that unilateral action leads to a drop in the overall level of Q and, hence, in carbon.

As before, this unilateral action can arise under a number of scenarios. First, a coalition of nations can institute a carbon tax on its members so as to tax the jointly produced global bad of carbon emissions. Such a tax has been

---

13. As in the last section, unilateral action may come from an alteration in coalition D's utility function.

recently proposed by the EC. Second, a single nation or group of nations can impose a standard or limit to its emissions. Some of the Scandinavian countries (e.g., Norway) have been considering this option. Third, the leftward shift of R_D to R_D' in figure 4 could result from a treaty between the coalitions in which only coalition D carries through. Once coalition D fulfills its pledge, coalition L would be best off by reneging on its promise to follow suit. For negatively sloped reaction paths, a Prisoner's Dilemma applies to treaty formation.

In figure 4, leader-follower behavior is shown to increase the overall level of Q and, hence, carbon emissions. If coalition D is the leader, while coalition L is the follower, then the leader-follower equilibrium occurs at S. Despite coalition L's cutback on q, the overall level of Q increases as compared with N.

As before, reduced cooperation among coalition D's members may improve the coalition's welfare. This reduced cooperation would imply a greater level of q^D for each level of q^L as compared with reaction path R_D, which assumes intracoalition cooperation. In consequence, the reaction path for coalition D would shift from R_D to the right and could increase the group's welfare if the new reaction path (not shown) were to intersect R_L between point N and the intersection point of curves II and R_L. A steep R_L and a small shift in R_D are conducive for the welfare improvement in D.

Neutrality does not apply in the presence of joint products,[14] because the existence of a private, coalition- or national-specific jointly derived outputs allows the distribution of income to have an impact on the overall level of Q. Thus, a role for aid from the developed nations is now present. A simple principle applies: aid must flow so that aid-assisted increases in carbon in the recipient nations are less than the reduction in carbon in the donor nations as real resources are transferred if other public outputs are ignored. Suppose that the developed nations transfer resources to the LDCs to preserve tropical forests. These resources will go to support development and, in so doing, will increase industrial emissions of carbon dioxide. One form of carbon emission will be substituted for another; net emissions may not fall. Will the world be better off? The answer depends on the mix of jointly produced public goods and bads that result when, say, deforestation is replaced with increased industrialization. Curbing deforestation not only reduces carbon emissions, but also provides global public goods of biodiversity and existence value. If these latter public goods are equal to or greater in value than the forgone economic activities in the donor countries, then a Pareto improvement will be achieved even when the overall level of carbon is unchanged. Hence, the inclusion of all joint products may justify an action that would otherwise be viewed as ineffective. If, moreover, the donors insist that the aid be used in energy-

---

14. This result was first established in Cornes and Sandler 1984.

efficient processes, then carbon levels may actually fall via the resource transfer.

Case 2: Positively Sloped Reaction Paths

The effects of unilateral action, treaties, and first-mover actions are much more favorable to all concerned when reaction paths are positively sloped. Strong complementarity among the jointly produced outputs may give rise to positively sloped reaction paths. If, for example, increased emissions of carbon create an increased demand for the jointly produced private output, then a complementarity exists. Increased global warming may, for instance, augment the demand for refrigerants such as CFCs, which, in turn, creates more global warming. Increased use of cement may also add to the demand for coolants, which, in turn, adds to more global warming. Thus, complementarity may be relevant for the jointly produced coalition-specific good and the global bad of carbon. Income effects may, however, be the most important rationale for positively sloped reaction paths. If the concern for global warming is income elastic and if, moreover, scientists portend dire consequences for increased global warming, then the income effect can be positive. Conventional wisdom supports the positive income elasticity of environmental concerns; hence, positively sloped reaction paths may be a reasonable outcome.

In figure 5, the initial Nash equilibrium is at point N, and the cross-hatched region comprises the points Pareto-superior to N. Since a portion of R_L lies within this region, unilateral action (i.e., the use of a per-unit tax combined with a lump-sum subsidy) on the part of coalition D can be Pareto improving; hence, a shift from R_D to R_D' leads to a new equilibrium at N' where both coalitions are better off. Additional action on behalf of coalition L to curb q^L and, hence, emissions of carbon would shift R_L to R_L' and further improve everyone's welfare at the new equilibrium T.

For positively sloped reaction paths, treaties between coalitions to reduce q can be self-enforcing, so that the Prisoner's Dilemma scenario does not apply. Once coalition D has fulfilled its pledge to reduce q^D for each level of q^L and N' is reached in figure 5, coalition L is better off to abide by the treaty, since its welfare level is higher at point T than at N'. The same is true for D if coalition L were to act first and achieve the new Nash equilibrium at the intersection of R_D and R_L'.

Noncooperative behavior within coalition D shifts its reaction path to the right and would increase emissions in D. Emissions in coalition L would also increase owing to the positive slope of the reaction path; hence, welfare would drop in both coalitions. Coalitions would, therefore, have an incentive *not* to dissemble.

Leader-follower behavior would also improve the welfare of both of the

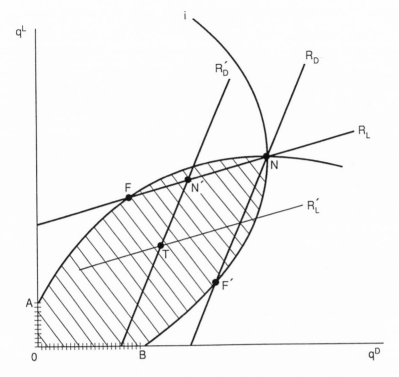

Fig. 5. Joint products and upward sloping reaction paths

coalitions, since the equilibrium would be between F and N when coalition D is the leader, or between F′ and N when coalition L is the leader. At the leader-follower equilibrium (not shown), both coalitions reduce their levels of q, so that the overall level of Q falls. The follower is apt to make the greatest reduction.

The Pareto path is AOB and does not consist of tangencies.[15] Along this path, one coalition's welfare increases as the other's welfare decreases. Cooperation is much more likely with positively sloped reaction paths, since reductions in q, and emissions, can be emulated by the other coalition. In fact, there is a self-enforcing incentive to augment cooperation until point T moves to Pareto path AOB.

Table 2 provides a summary of the primary results. The analysis here may also be applied to other commons problems. For example, arms race

---

15. In a pure public good model of opposing alliances, this type of Pareto contract curve was first noted by Bruce 1990.

TABLE 2. Summary of Results for Joint Product Model of Carbon Emissions

Cases[a]	Emissions in D	Welfare in D	Emissions in L	Welfare in L	Total Emissions
Coalition D is	increase	increase	decrease	decrease	increase
Stackelberg leader	(decrease)	(increase)	(decrease)	(increase)	(decrease)
Unilateral action[b] by	decrease	decrease[c]	increase	increase	decrease
coalition D	(decrease)	(increase)	(decrease)	(increase)	(decrease)
Noncooperative behavior	increase	decrease	decrease	decrease	increase
within coalition D	(increase)	(decrease)	(increase)	(decrease)	(increase)

[a]Results for negatively sloped reaction paths are listed in the top row of each scenario, while results for positively sloped reaction paths are listed *in parentheses* in the second row of each scenario.

[b]Includes the case where coalition D acts first to fulfill a treaty pledging reduced emission levels.

[c]Welfare change for coalition D is ambiguous if the unilateral action involves a change in its preferences or an increase in the price of the activity that is not offset by a lump-sum subsidy.

interactions may give rise to positively sloped reaction paths so that unilateral disarmament might be desirable. This may have occurred recently between the United States and the Commonwealth of Independent States regarding arms reduction.

Concluding Remarks

In many ways, transfrontier commons problems share the features of the classic problem of the voluntary provision of a public good. Nevertheless, important distinctions exist. In particular, joint products tend to be associated with transfrontier pollution, and often involve a nation-specific private output and a global public bad. Additional public goods and bads may also be included. The existence of these joint products yields some important results. First, a role for redistributive income policy can exist, since the neutrality theorem, which inhibits income policy in the case of pure public goods, does not apply. Second, the degree of suboptimality may be attenuated as compared with the pure public good scenario. Third, a number of different cases exist, including those in which unilateral action and leader-follower behavior can be self-enforcing and beneficial to all involved. Moreover, treaties between coalitions may be self-enforcing due to the underlying game structure.

The next logical step is to extend the analysis from noncooperative to cooperative games. Since the no-agreement payoffs revert to the Nash equilibrium of the noncooperative game, the noncooperative game analysis is a necessary first step. Another direction is to examine the formation of the coalitions D and L, since collective action problems may be more problematic at this level. Finally, a dynamic noncooperative analysis represents another important research direction.

The current chapter shares similarities and differences when compared with the Kleinblittersdorf case study in the Feld, Pommerehne, and Hart (hereafter FPH) chapter in this volume. This FPH chapter concerns the actions of citizens of Kleinblittersdorf to bribe residents of a neighboring French community of Grosbliederstroff not to build a waste treatment plant. In both chapters, collective action aspects are stressed. FPH focus on interactions *within* a community to raise contributions for a public good, while I focus on interactions between collectives of nations to provide a public good. Both chapters bring up notions of agent-specific private benefits that are jointly produced with the public good. FPH call these private aspects "selective incentives." I emphasize noncooperative game concerns, whereas FPH stress Coase-like cooperative game concerns (e.g., bargaining). FPH analyze contributions in terms of empirical testing; the current chapter only investigates the theoretical implications. Although both chapters address the free-rider issue, the current chapter studies unilateral actions, neutrality, and alternative strategic assumptions (leader-follower and Nash). Both chapters conclude that collective action is feasible under the right conditions.

REFERENCES

Barrett, S. 1991. "The Problem of Global Environmental Protection." In D. Helm, ed., *Economic Policy towards the Environment*. Oxford: Blackwell.

Bergstrom, T. C.; L. Blume; and H. Varian. 1986. "On the Private Provision of Public Goods." *Journal of Public Economics* 29, no. 1:25–49.

Bruce, N. 1990. "Defense Expenditures by Countries in Allied and Adversarial Relationships." *Defence Economics* 1, no 3:179–95.

Carraro, C., and D. Siniscalco. 1992. "The International Dimension of Environmental Policy." *European Economic Review* 36, nos. 2–3:379–87.

Cornes, R., and T. Sandler. 1984. "Easy Riders, Joint Production, and Public Goods." *Economic Journal* 94, no. 3:580–98.

———. 1986. *The Theory of Externalities, Public Goods, and Club Goods*. New York: Cambridge University Press.

———. 1994. "The Comparative Static Properties of the Impure Public Good Model." *Journal of Public Economics* 56, no 3:257–72.

Hoel, M. 1991. "Global Environmental Problems: The Effects of Unilateral Actions Taken by One Country." *Journal of Environmental Economics and Management* 20, no. 1:55–70.

Ko, I.-D.; H. E. Lapan; and T. Sandler. 1992. "Controlling Stock Externalities: Flexible versus Inflexible Pigovian Corrections." *European Economic Review* 36, no. 6:1263–76.

Mäler, K.-G. 1991. "International Environmental Problems." In D. Helm, ed., *Economic Policy towards the Environment*. Oxford: Blackwell.

Morgenstern, R. 1991. "Towards a Comprehensive Approach to Global Climate Change Mitigation." *American Economic Review* 81, no. 2:140–45.

Morrisette, P. M., and A. J. Plantinga. 1991. "The Global Warming Issue: Viewpoints of Different Countries." *Resources* 103, no. 1:2–6.

Nordhaus, W. D. 1991. "A Sketch of the Economics of the Greenhouse Effect." *American Economic Review* 81, no. 2:146–50.

Sandler, T. 1992. *Collection Action: Theory and Applications*. Ann Arbor: University of Michigan Press.

Warr, P. G. 1983. "The Private Provision of a Public Good Is Independent of the Distribution of Income." *Economics Letters* 13, no. 2:207–11.

White, R. M. 1990. "The Great Climate Debate." *Scientific American* 263, no. 1:36–43.

World Resources Institute. 1990. *World Resources 1990–91: A Guide to the Global Environment*. New York: Oxford University Press.

CHAPTER 12

Political Institutions and Pollution Control

Roger D. Congleton

The essential economics of pollution are fairly straightforward. Pollution is largely an undesired by-product of production and transport. Consequently, countries that engage in more production and/or more transport will, *ceteris paribus*, generate more pollution. A country's output of pollution, thus, is evidence of its material prosperity, and is to a large extent a consequence of purely economic considerations: the pool and distribution of natural resources available to it, the productivity of its people, the nature of available production technologies and market structure.[1]

However, clearly *more* than economic considerations are involved. Political institutions affect the extent of a nation's output of effluents insofar as they directly affect environmental policies, or indirectly encourage or discourage economic development. Grier and Tullock (1989) demonstrate that economic growth rates in nations with liberal democratic institutions are higher than those with less liberal political arrangements, other things being equal. Ignoring direct regulation, democratic regimes would tend to exhibit higher growth rates of pollutants than less liberal regimes. Democratic political institutions will moderate these economic tendencies if they are systematically more likely to yield polices that encourage firms and/or consumers to adopt less polluting technologies.

Domestic political institutions also affect efforts to negotiate international environmental standards insofar as decisions to sign and abide by

Reprinted from *Review of Economics and Statistics*, 1992, 412–21. Financial support from the International Institute is gratefully acknowledged. The paper has benefited from discussions with John Moore, the data assistance of Nicole Verrier, and the comments of three anonymous referees.

1. See Baumol and Oates 1988 for an overview of economic aspects of pollution. Tobey (1990) and Leonard (1988) discuss trade aspects of environmental policies. Tobey, however, was unable to find a relationship between domestic environmental policies and patterns of international trade. For recent overviews of the economics of air pollution see Firor 1990. Dryzek (1987) discusses the effects of political institutions on environmental policy. The analysis developed here supports Dryzek's contention that liberal democracies are more responsive to environmental concerns than authoritarian or bureaucratic regimes.

multilateral agreements are ultimately decisions of distinct polities. In most cases, states that find a particular proposed regulation or convention to be against "their interests" can opt out of the negotiations and ignore any conventions signed by those that remain at little cost; as there are, at best, only weak international legislative and enforcement institutions. Inasmuch as domestic political institutions effectively determine the range of interests that are accounted for in policy decisions, regimes that enact stringent domestic environmental standards are likely to find international agreements along similar lines to be in their interest.[2]

The purpose of this chapter is to explore the effect of political institutions on the willingness of governments to control outputs of domestic effluents. To this end, models of democratic and authoritarian environmental policy making are developed in section 1. The analysis demonstrates that an authoritarian confronts a higher relative price for pollution abatement than a median voter does. In cases where this relative price effect dominates, authoritarian regimes will adopt less stringent domestic environmental standards than democratic regimes, and by extension, be less willing to sign international conventions on environmental matters.

Empirical evidence on the latter proposition is developed in section 2 of the chapter. Generally, the results support the contention that liberal democracies are both relatively high sources of air pollution (because of their higher income) and more likely to sign international conventions on the environment. Section 3 summarizes the results and suggests possible extensions.

1. Environmental Politics

The essential difference between authoritarian regimes and democracies is their decision-making procedure. Democracies make policy by counting the votes of ordinary citizens or their elected representatives. Authoritarian regimes only take account of the "votes" of an unelected elite, in the limit the "vote" of a single ruler. While many secondary features of authoritarian and democratic regimes indirectly influence public policies insofar as they affect public debate or planning horizon, they are initially neglected to focus attention on similarities in decision-making procedures. With this in mind, global markets in technology and capital are assumed to be sufficiently competitive that technological aspects of environmental policy are the same for authori-

---

2. Economic aspects of domestic and international environmental protection differ somewhat for import and export industries. For example, export industries restricted by environmental domestic standards would be inclined to oppose cost-increasing domestic standards. However, once domestic standards are enacted, affected domestic industries would tend to support similar strictures for their international competitors.

tarian and democratic regimes.[3] Voters and dictators are assumed to face the same technological tradeoff between environmental quality and nonenvironmental income.

At this level of abstraction, both authoritarian and democratic decision-making procedures are fundamentally similar in that they yield environmental policies that maximize the welfare of a single individual given various political and economic constraints. Under a democratic regime, the pivotal voter is the mean or median voter.[4] Under an authoritarian regime, the pivotal decision maker is the dictator (or the pivotal member of the ruling group). Any systematic difference between policies adopted by these regimes must be ascribed to differences in the parameters of their decision problems.

A. Utility-Maximizing Pollution Standards

For purposes of analysis, individuals are assumed to maximize a two-dimensional utility function defined over measured real income (or consumption) as per GNP accounting practices and environmental quality. Measured real income is a pure private good measured as an index of goods purchased in markets. Environmental quality is a public good measured as an index of the average

---

3. Nations will face different costs for capital in perfectly competitive markets if they are more or less stable and/or have different regulatory environments. However, such differences are fundamentally political in nature inasmuch as they reflect political uncertainties and/or import-export restrictions. Kormendi and Meguire (1985) report that political regimes provide the strongest explanation of international investment flows. Costs will also vary if there are important nontransportable natural resources that affect local relative prices of alternative production techniques. Cost differences due to differences in tax policies, infrastructure, land use restrictions, and so forth can be modeled in a manner similar to that used for economic standards developed above.

The chapter models the enactment of environmental standards rather than pollution taxes or land use regulations to simplify analysis. Buchanan and Tullock (1975) argue that political considerations tend to make standards more attractive than appropriate Pigovian taxes.

4. Stochastic models of voter choice generally imply an equilibrium at the average voter's ideal point. See Enelow and Hinich 1984 or Coughlin 1992 for an overview of such models. In the case where voter ideal points are symmetrically distributed, the mean and median voter positions are the same and the policy predictions of both models are identical. The median voter model is used in this paper because of its more conventional representation of voter decision making and because median voter models are more widely used in the public policy literature. Sufficient conditions are met in the model for the existence of a median voter equilibrium. The policy domain is single dimensioned and the concavity of equation (2) implies that objective functions are single peaked in this domain.

The political effects of economic interest groups are ignored in this paper. As a first approximation, the efforts of economic interest groups are assumed to offset each other as in the symmetric case characterized by Congleton (1989), leaving the median voter or authoritarian decisive.

density of undesired (often toxic) chemicals in the voter's environment. These two areas of choice, while technologically linked, are disjoint since environmental quality is not normally included in measured real income.

The relationship between measured (real) national income and the measured (real) income of a typical individual is assumed to be a monotone increasing function of national income. This is a fairly severe assumption since environmental regulations often have quite different effects on different industries or regions within a country. However, the voter of most interest for the purposes of this paper is the median voter, who, because of his position at the midpoint of the distribution of voter ideal pollution standards, is unlikely to be one of the most (or least) affected individuals. Authoritarians also have incomes that are linked to the national economy insofar as their income is based on tax revenues and gratuities. Initially, each individual is assumed to receive a constant share of GNP. Given national income Y and an individual's income share v, personal income is $C = vY$.

National income is a function of environmental regulations, E^*, market institutions, M, and resource base, R, in the country of interest. National income increases as economic resources increase and as market arrangements become less centrally managed. National income initially rises as environmental standards become more stringent, peaks at standard E^y, and falls thereafter. Over the initial range, more stringent environmental standards increase national output by improving the health and productivity of labor and/or by freeing resources previously devoted by individuals to reduce exposure to the local environment, air conditioners and the like, for more valuable uses. Over the latter range, more stringent environmental standards reduce ordinary national output as less productive technologies are mandated and inputs are diverted from ordinary economic production to environmental improvement without offsetting productivity improvements.[5] In this range, more stringent regulations improve average environmental quality but increase the cost of consumer goods relative to income, reducing measured national income and thereby the measured income of individuals.

---

5. For those skeptical of the initial range where GNP increases with increases in environmental controls, recall that resources have been invested in waste management (dumps, latrines, and so forth) even by prehistoric human communities. These investments tend to improve the health and longevity of community members, with a consequent increase in the effective labor supply of the community. This suggests that over some range environmental standards increase productivity and thereby personal income. Ridgeway (1970) reports that such arguments played an important role in promoting sanitation efforts by Western democracies during the last century.

To see that a similar upward-sloping range exists for air quality, imagine the problem faced by a mine owner. If he fails to maintain some minimal level of environmental controls within the mine, his labor force will be less productive as air quality and temperature take their toll. A suitable investment in ventilation and cooling equipment can increase productivity.

Economic regulations affect the size of a voter's opportunity set through relative and absolute price effects that affect the size of national GNP. In this respect environmental standards (and other regulations) are unlike ordinary public expenditure programs in which tax revenues are raised to finance provision of a publicly produced service. The cost of environmental quality is not reflected by changes in ordinary tax burden, but rather by indirect effects that environmental standards have on personal income.

The link between environmental standard E^* and environmental quality is assumed to be probabilistic. This reflects stochastic elements of the underlying natural processes and scientific uncertainty about the physical and social mechanisms involved. An individual's assessment of the probability of environmental deterioration, $P = p(E|E^*, Y)$, falls as the environmental standard, E^*, becomes more stringent and increases as national output, Y, increases. Environmental standards are a form of social insurance which reduce downside environmental risk.

Each individual prefers the environmental standard that maximizes lifetime expected utility given various personal constraints. Given, our characterization of the choice at issue and a finite time horizon, $T,$[6] an individual's preferred standard can be modeled as that which maximizes

$$U^e = \int_0^T \left[U^o P^o + \int u(C, E, t) P(E \mid E^*, Y, t) \, dE \right] dt \qquad (1.0)$$

where $U^o \equiv U(Y, E^o, t)$ and $P^o \equiv 1 - \int P(E \mid E^*, Y, t) \, dE$

with $U_C > 0, U_E > 0, U_t < 0, U_{CC} < 0, U_{CE} > 0, U_{EE} < 0$

$U_{tC} < 0, U_{tE} < 0, U_{tt} > 0, P \geq 0, P_{E^*} < 0, P_{Y/A} > 0,$

subject to

$$C = vY \qquad (1.2)$$

$$Y = y(E^*, M, R, t) \qquad (1.3)$$

with $Y_R > 0, Y_{E^*} > 0$ and $Y_{E^*E^*} > 0$ for $E^* < E^y,$

$Y_{E^*} = 0$ for $E^* = E^y,$ and $Y_{E^*} < 0$ for $E^* > E^y.$

6. Congleton and Shughart (1990) provide evidence that median years to retirement and expected longevity affect policy decisions regarding social security benefit streams. Their results suggest that the median voter has a finite planning horizon in excess of thirty years.

Subscripted variables represent partial derivatives with respect to the variable subscripted. U^o is utility generated by the original (or maximum) level of environmental quality with a measured income determined by the implied public policy. Conditional probability function P represents the probability distribution of environmental qualities below E^o in period t. Thus, the integral of P represents the probability that environmental quality falls below E^o. Utility increases as measured income, C, and/or environmental quality, E, increase and declines with postponement, t.

Substituting the constraints into the objective function, in order to specify the choice in terms of the environmental standard, E^*, yields:

$$U^e = \int_0^T \left\{ U(vY, E^o, t) \, P^o \right.$$

$$\left. + \left[\int u(vY, E, t) \, P(E|E^*, y(E^*), t) \right] dE \right\} dt \qquad (2)$$

Differentiating with respect to environmental standard E^*, and setting the result equal to zero yields a single first order condition for a typical individual's ideal environmental standard:

$$\int_0^T vU_C^o \, Y_{E^*} \, P^o + U^o \, (P_{E^*}^o + P_Y^o \, Y_{E^*})$$

$$+ \left[\int vU_C \, Y_{E^*} \, P + U \, (P_{E^*} + P_Y \, Y_{E^*}) \, dE \right] dt = 0 \qquad (3.0)$$

or

$$\int_0^T vU_C^o \, Y_{E^*} \, P^o - \left[\int vU_C \, Y_{E^*} \, P \, dE \right] dt$$

$$= \int_0^T - U^o \, (P_{E^*}^o + P_Y^o \, Y_{E^*}) + \left[\int U \, (P_{E^*} + P_Y \, Y_{E^*}) \, dE \right] dt. \qquad (3.1)$$

Equation (3.1) demonstrates that the effect of environmental regulation on utility occurs through its effects on personal income and the probability distribution of environmental quality. Each individual prefers the environmental standard that sets the expected present value of his subjective marginal cost for the standard in terms of reduced measured income (consumption) equal to

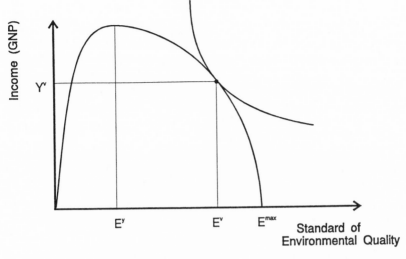

Fig. 1

the present discounted value of the time stream of marginal utility from greater environmental quality.

Such environmental standards have the geometry of the curves in figure 1. The tradeoff between economic output and environmental quality is the envelope of the opportunity set. It represents the steady-state tradeoff between measured (or pecuniary) income and environmental quality. The indifference curves represent constant levels of expected lifetime utility over the time horizon of interest. Note that the relevant part of the constraint is from E^y to $E^{\max}$. A rational decision maker who regards neither economic income nor environmental quality as "bads" would never intentionally adopt a standard below E^y. An individual who values income but cares nothing for the environment, *per se,* prefers the standard which maximizes measured income, E^y. The median voter is by definition in the middle of the distribution voter ideal points, and therefore opts for an intermediate solution similar to that depicted. The situation of a dictator is not similarly constrained, and thus more extreme environmental policies are possible. Yet the general shape of the opportunity set implies that both sorts of regimes tend to have some environmental regulation.

The implicit function theorem implies the existence of an equation, based on (3.0), which represents an individual's preferred environmental standard as a function of variables beyond his control.

$$E^{**} = e(v, T, R, M) \tag{4}$$

Ultimately, the level of environmental regulation preferred depends upon the individual's share of national income, v, his time horizon, T, the resource base of the country, R, and its market institutions, M. The implicit function differentiation rule allows the effect of changes in these parametric variables on environmental standards to be calculated.

Define the left-hand side of equation (3.0) to be H, then

$$E_v^{**} = H_v/-H_E \tag{5.1}$$

$$E_T^{**} = H_T/-H_E \tag{5.2}$$

$$E_R^{**} = H_R/-H_E \tag{5.3}$$

$$E_M^{**} = H_M/-H_E. \tag{5.4}$$

H_E is the second derivative of equation (2) with respect to environmental quality, which must be negative at the expected utility maximum. Consequently, the signs of equations (5.1) through (5.4) are determined by the numerators. Somewhat surprisingly, given the simplicity of the model used, none of these are unambiguously determined. An increase in the share of national income going to the individual, an increase in his time horizon, or an increase in national resources or geographical area can lead to stronger or weaker environmental standards according to the relative sizes of various interaction terms.

An increase in the fraction of national income going to the individual of interest increases the marginal cost of environmental standards faced by him, since he will now bear a larger fraction of associated reductions in national income. On the other hand, his income is higher for any given standard and thus would, except for the relative price effect, opt for greater expected environmental quality. In the case where the relative price effect dominates, $E_v^{**} < 0$, and the utility maximizing environmental standard become less stringent as the individual's share of national income increases.

The effect of an increased planning horizon depends on the incremental costs and benefits from higher environmental standards at the end of the period evaluated. It is often argued that the time stream of benefits from environmental standards is such that the costs of environmental standards are concentrated in the early periods while the benefits are concentrated in later time periods. See, for example, Nordhaus 1989. Consequently, it is likely that at the end of the planning horizon the marginal benefits of environmental standards exceed their marginal costs which implies that $E_T^{**} > 0$. In this case, the longer the time horizon used, the more stringent an individual's ideal environmental standards tend to be.

The effect of changes in a nation's resource base or market institutions is more complex. A larger resource base or more efficient collection of market institutions implies a larger national income for any environmental standard, which increases the probability of environmental deterioration and makes more stringent environmental standards more attractive. However, greater national income also implies that each voter is wealthier and inclined to "purchase" both more measured income and more environmental quality. On the other hand, greater personal wealth may increase the marginal cost of environmental regulation to the voter. In the case where increased environmental downside risk and income effects more than offset any increase in the marginal cost of environmental standards, equations (5.3) and (5.4) will exceed zero. Here, the stringency of environmental standards would increase as the resource base and/or quality of market institutions increase, $E_R^{**} > 0$ and $E_M^{**} > 0$.

B. Differences among Utility-Maximizing Political Regimes

The preceding analysis implies that utility-maximizing decision makers will disagree about the optimal environmental standard, even if their utility functions are fundamentally similar, if their positions in the national economy, planning horizon, or national endowments differ. To demonstrate that political regimes affect environmental policies, it is sufficient to show that the circumstances of policy makers under these two systems tend to be systematically different.

The effect of a change in regime on environmental policy can be analyzed with reference to equations (5.1) and (5.2). Recall that the median voter is approximately the voter with the median income share and time horizon. Authoritarians have greater than the median income share and probably tend to have a shorter than average time horizon given the high turnover of authoritarian regimes.[7] Under the restricted circumstances previously discussed, a larger national income *share* and shorter time horizon tend to reduce the stringency of the desired environmental standard. Consequently, authoritarians tend to prefer a lower environmental standard than a median voter does.

The essential geometry of such median voter and authoritarian choices is depicted in figure 2. The effect of a change from authoritarian regimes to democratic regimes in a given economy is analogous to a rise in the price

7. The median term of office for African authoritarians reported by Bienen and van de Walle (1989) is about four years. Differences in utility functions are also very likely. The highly uncertain career path to the top of an authoritarian regime and the relatively short typical term of office suggest that authoritarians tend to be relatively less risk averse than median voters tend to be. The relatively risky nature of the ascent to power suggests that authoritarians will be less inclined to purchase insurance of any kind, including environmental insurance.

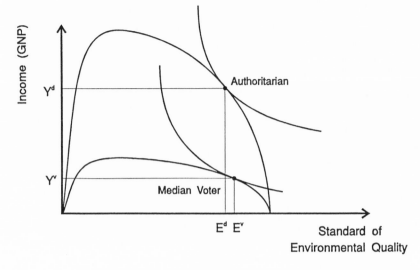

Fig. 2

of environmental quality. The higher marginal cost faced by a dictator is evidenced by the more steeply downward sloping portion of his income/ environmental-standard constraint. In the case depicted, the marginal rate of substitution between income and environmental quality is independent of measured income, and consequently differences in marginal cost are decisive. Here, the authoritarian opts for more pecuniary income and less stringent environmental standards than the median voter does.

2. Empirical Evidence

The preceding analysis makes a fairly good case for democratic regimes to be inclined to adopt more stringent environmental regulations than authoritarian regimes, although the results are not completely general. The model provides a framework for testing this and other hypotheses generated in the course of analysis. Under both authoritarian and democratic regimes, pollution standards are functions of national endowments and characteristics of the decisive voter. Equation (4) is the reduced-form equation for environmental policy formulation. If democratic regimes are more inclined to enact strict environmental standards than authoritarian regimes, then estimates of equation (4) should have positive coefficients for democratic variables. Reduced-form equations for other endogenous variables are similarly functions of policy-maker characteristics, national resources, and market institutions. The model also characterizes several structural equations. The estimates reported in this section are broadly consistent with the model, although a variety of data problems had to be circumvented in order to generate the estimates.

First, data on domestic environmental regulations tend to be sparse and the regulations themselves different enough to make a cross-section analysis problematic. On the other hand, there have been several recent international agreements on environmental concerns. So while it is not possible to test hypotheses with respect to domestic policies, it is possible to test whether a country's political regime affects its international environmental policies. To the extent that domestic considerations dictate whether a nation signs on to a particular covenant or protocol, these international agreements also indirectly cast light on domestic policies.

In 1985, a global convention was negotiated under United Nations auspices at Vienna which committed signatories to coordinate research efforts and to enact domestic legislation for reducing emissions of ozone depleting substances, chiefly chlorofluorocarbons (CFCs). In 1987, a protocol was developed in Montreal which obliged developed countries to reduce their CFC production and consumption levels to half their 1986 levels by June 1999 and developing nations to limit their future consumption of CFCs.[8] In both agreements, countries could exempt themselves from the negotiated targets by not signing the final document. As of 1989, 27 countries had signed the 1985 conventions or enacted national legislation which accomplished the same end. Twenty-nine countries signed or enacted national legislation supporting the Montreal protocol (World Resources Institute 1989, table 24.1). Nineteen countries supported both agreements. To the extent that domestic politics determines whether a particular country agrees to sign a particular covenant or protocol, my analysis suggests that democratic regimes will be more likely to participate in these international agreements than other regimes, other things being equal.

---

8. Paragraph 4 of Article 2 of the Protocol specifies that "each party shall ensure that for the period 1 July 1998 to 30 June 1999, and in each twelve month period thereafter, its calculated level of consumption on the controlled substances in Group I of Annex A (chiefly CFCs) does not exceed, annually, fifty per cent of its calculated level of consumption in 1986." Paragraph 1 of Article 5 exempts developing countries with "consumption of controlled substances . . . less than 0.3 kilograms per capita" from reductions in their current consumption levels but restricts them to "either the average of its annual calculated level of consumption for the period 1995 to 1997 inclusive or a calculated level of consumption of 0.3 kilograms per capita, which ever is lower." CFC levels ranged from 0 to 0.91 kilograms per capita in 1986 within the 118 countries used above. A typical Western democracy had levels of 0.75 kilograms per capita.

Signatories or contracting parties to the Montreal Protocol, as of 1989, include Egypt, Ghana, Kenya, Morocco, Senegal, Togo, Canada, Mexico, Panama, United States, Venezuela, Israel, Japan, Belgium, Denmark, Finland, France, West Germany, Greece, Italy, Luxembourg, Netherlands, Norway, Portugal, Sweden, Switzerland, United Kingdom, USSR, and New Zealand (World Resources Institute 1989, table 24.1).

Second, data on the personal characteristics of median voters and dictators are largely unavailable. However, if these characteristics vary systematically with the type of regime, one can use regime type as a proxy for the personal characteristics of decision makers under the two regimes. Third, classification of countries into democratic and authoritarian regimes is also somewhat problematic. Many authoritarian regimes have formal institutional structures that superficially parallel those of democratic regimes. For example, according to the *CIA Fact Book* (1988), both Zaire and Syria have political parties and elections, although most would consider these governments to be authoritarian. Gastil (1987, pp. 40–41) divides countries into seven categories according to their political liberties. Although Gastil's classification scheme yields a plausible ranking of countries, the extreme values are used here to distinguish democratic from nondemocratic government types. Countries that receive the highest values for political liberties (categories 1 and 2, table 3) are essentially well-functioning democracies whose citizens have broad political rights of participation. Differences in time horizon and share of national income relevant for policy makers are proxied by the resulting 0–1 dichotomous variable labeled *Democratic Country* in the tables that follow. Gastil also constructed a ninefold classification of countries according to market structure (1987, p. 74, table 8). The two most market-oriented categories are countries with relatively unregulated open markets. A 0–1 dichotomous variable constructed from these two categories, denoted *Capitalist Country* in tables 1 and 2, is used as a proxy for the efficacy of a country's market institutions.

Fourth, the model implies that a country's resource base plays a role in determining national standards insofar as resources affect personal income and the marginal cost of environmental regulation. However, reliable data on a nation's mineral reserves and labor force are available for relatively few countries. For example, complete data on recoverable fossil fuel reserves are available from the World Petroleum Institute for only 46 countries. However, if resource endowments are largely the result of geological accidents more or less uniformly distributed about the world, one can expand the sample by using national area to proxy the national resource base. (Area is highly correlated with proven fossil fuel reserves.) Labor force numbers are similarly problematic and are proxied by national population. Data for population and area are from the *World Fact Book* (CIA, 1988). Data on real GNP in U.S. dollars (1987) are from the World Petroleum Institute (1990/1991).

Linear estimates of equation (4) are reported in the first four columns of table 1. Columns 1 and 2 report logit estimates of the probability that a particular country would have agreed to the 1985 Vienna Convention or the 1987 Montreal Protocol, using proven reserves of fossil fuels as a proxy for national resource endowment. Columns 3 and 4 report similar estimates based

TABLE 1. Estimates of the Probability of Signing Environmental Treaties

Variable Name	CFC85$_{89}$ Logit	CFC87$_{89}$ Logit	CFC85$_{89}$ Logit	CFC87$_{89}$ Logit	CFC85$_{89}$ Logit	CFC87$_{89}$ Logit
C	-0.437	-1.869	-2.254	-2.114	-2.448	-2.286
	(0.642)	(2.330)*	(5.516)***	(5.37)***	(5.586)***	(5.474)***
Democratic country	2.228	3.033	2.596	2.836	2.227	2.506
	(2.863)**	(3.422)***	(4.836)***	(5.151)***	(3.832)***	(4.261)***
Capitalist country	-0.896	-0.755	-0.267	-0.726	-0.251	-0.748
	(1.049)	(0.836)	(0.458)	(1.194)	(0.415)	(1.193)
GNP per capita					0.064	0.058
					(1.817)*	(1.675)*
Oil reserves	-6.05 E-7	-1.040 E-5				
	(0.050)	(0.556)				
Coal reserves	2.252 E-5	1.403 E-6				
	(1.706)*	(0.157)				
Gas reserves	1.490 E-8	2.423 E-6				
	(0.628)	(0.963)				
Area			2.356 E-4	1.102 E-4	2.356 E-4	9.672 E-5
			(1.857)*	(1.266)	(1.672)*	(1.097)
Population	-0.017	-0.00202	-3.624 E-3	-8.270 E-4	-0.004	-0.0007
	(1.868)*	(0.355)	(0.784)	(0.324)	(0.671)	(0.251)
Sample size	46	46	118	118	118	118
Log likelihood	-22.671	-20.830	-46.377	-47.264	-44.632	-45.814

*Significant at the 10 percent level.
**Significant at the 1 percent level.
***Significant at the 0.1 percent level.

on the full 118-country sample using national area as a proxy for national resources. Columns 5 and 6 augment the model by adding per capita income to the list of dependent variables. This creates some risk of simultaneous equation bias but demonstrates that the results of the reduced-form estimates of the model are not an artifact of the higher income levels of democratic governments.

Note that, consistent with the main thesis of this chapter, the coefficient for the democratic country variable is significantly different from zero at the one-thousandth level of significance for each of the six estimates of international environmental policy, although the economic endowment and market structure variables are not. This suggests that environmental policy decisions are largely determined by a nation's political institutions rather than by its economic resource endowments. Moreover, these results are robust in the sense that small changes in data set or specification do not change the signs or significance of the coefficient estimates.

Table 2 reports estimates of other structural and reduced-form equations from the model. Columns 1–4 focus on two pollutants of current interest because of their roles as greenhouse gases: net methane and CFC output as reported in *World Resources 1989–1990*. Column 5 reports a reduced-form estimate of real gross national product. Column 1 reports a two-stage estimate of methane output. This is an estimate of the probability function of environmental quality characterized used in equation (1) of the model. Recall that observed environmental quality is a probabilistic function of RGNP and environmental standards. A country's participation in the Montreal Protocol is used as a proxy for its regulatory environment. Both coefficient estimates are statistically distinguishable from zero. An increase in a country's GNP increases its methane output while the domestic propensity to regulate reduces it. Column 2 provides indirect evidence of the effects of domestic regulation on national outputs of methane per unit of GNP. Democratic regimes produce more methane in total but significantly less per unit of national output. Together with the other results, this suggests that liberal democracies are more inclined to regulate environmental outputs than other regimes are.

Columns 3 and 4 are reduced-form estimates of net outputs of methane and CFCs. A country's area again serves as a proxy for its natural resource base. The coefficients all have positive signs consistent with our previous discussion. However, again, only the coefficient estimates for the effects of liberal democratic regimes and national area are statistically significant. Democratic regimes produce more methane and more CFCs than their less liberal and/or more authoritarian counterparts. This apparently reflects the higher national incomes associated with liberal democracies and the consequent production of more refuse of all sorts.

Both series of estimates are broadly similar. They suggest that liberal

TABLE 2. Estimates of National Effluents and Economic Production

Variable Name	Methane 2-Stage LS	Methane/GNP OLS	Methane OLS	CFC OLS	GNP OLS
C	235.204 (2.079)*	12.363 (6.553)***	-86.239 (1.367)	-2.105 (0.849)	-73.640 (1.421)
GNP_{1987} (in U.S. dollars)	2.481 (8.145)***				
Montreal signatory or contracting party	-1011.481 (2.506)**				
Democratic country		-9.501 (2.980)**	236.868 (2.904)**	12.285 (2.934)**	254.147 (2.904)**
Capitalist country		0.128 (0.042)	122.570 (1.483)	4.103 (1.038)	122.570 (1.483)
Area		-0.0028 (0.478)	0.1004 (6.234)***	0.0043 (5.601)***	0.1004 (6.234)***
Population		0.00885 (0.720)	0.161 (0.478)	0.008 (0.499)	0.1612 (0.477)
Sample size	118	118	118	118	118
R-square	.219	0.082	0.748	0.321	0.386
Log likelihood	-964.424	-483.430	-897.538	-515.596	-874.302
F-statistic	16.092	2.522	84.087	14.829	17.747

*Significant at the 10 percent level.
**Significant at the 1 percent level.
***Significant at the 0.1 percent level.

democracies are significantly more likely than other regimes to have supported global efforts at environmental regulation. While the signs of other coefficients are plausible, for example, capitalist countries are less inclined to regulate than other regimes, the results cannot rule out the possibility that the other variables had no effect on the probability of supporting the Vienna Convention or the Montreal Protocol. In contrast, the coefficients for democratic regime are statistically significant in every case.[9]

3. Conclusion

This chapter has argued that environmental policies are affected by political institutions. Analysis of incentives faced by authoritarian regimes and democratic policy makers implies that relevant decision makers in democracies have a smaller marginal cost for pollution control than authoritarians do. Moreover, the highly uncertain career path to the top of an authoritarian regime and their relatively short term of office suggest that authoritarians tend to have a relatively shorter time horizon and be relatively less risk averse than median voters tend to be. Together these differences imply that authoritarian regimes are inclined to enact less stringent environmental standards than democratic regimes.

Empirical evidence supports the contention that political institutions affect domestic and international environmental policies. Cross-sectional analysis of both pollution outputs and willingness to take part in international conventions on the environment strongly suggest that liberal democracies are more willing to regulate environmental effluents than less liberal regimes. The empirical results support the contention that political institutions largely determine environmental regulation, rather than technological aspects of pollution control or market structure.

The results also suggest that international agreements on environmental matters of global concern will attract more signatories as the number of democratic regimes increases. Seen in this light, the recent increase in the number of countries with democratic political institutions implies that global environmental agreements will be more broadly supported in the future.

---

9. Similar results for domestic policies can be obtained using indexes from Gastil 1987, and Walter and Ugelow 1979. Walter and Ugelow constructed a seven-value index of the severity of environmental regulations for 25 countries. Regressing this environmental index (E) on the seven-value Gastil index (adjusted to make higher numbers imply greater political liberty) of the liberalness of a country's political institutions (D) yields

$$E = 2.135 + 0.390D \qquad F = 4.391.$$
$$(1.86)^* \quad (2.07)^*$$

In this quite restricted sample, the more liberal (democratic) a political regime is, the more severe its environmental policies tend to be.

REFERENCES

Baumol, W., and Oates, W. 1988. *The Theory of Environmental Policy*. 2d ed. New York: Cambridge University Press.

Bienen, H., and van de Walle, N. 1989. "Time and Power in Africa." *American Political Science Review* 83:19–34.

Buchanan, J. M., and Tullock, G. 1975. "Polluters' Profits and Political Response: Direct Controls versus Taxes." *American Economic Review* 65:139–47.

Congleton, R. D. 1989. "Campaign Finances and Political Platforms: The Economics of Political Controversy." *Public Choice* 62:101–18.

Congleton, R. D., and Shughart, W. F. 1990. "The Growth of Social Security: Electoral Push or Political Pull?" *Economic Inquiry* 28:109–32.

Coughlin, P. 1992. "Majority Rule and Elections Models." *Journal of Economic Surveys* 4:157–88.

Dryzek, J. S. 1987. *Rational Ecology Environment and Political Economy*. New York: Basil Blackwell.

Enelow, J. M., and Hinich, M. J. 1984. *The Spatial Theory of Voting*. New York: Cambridge University Press.

Firor, J. 1990. "The Straight Story about the Green House Effect." *Contemporary Policy Issues* 8:3–15.

Gastil, R. D. 1987. *Freedom in the World: Political Rights and Liberties 1986–1987*. New York: Greenwood Press.

Grier, K. B., and Tullock, G. 1989. "An Empirical Analysis of Cross-National Economic Growth, 1951–80." *Journal of Monetary Economics* 24:259–76.

Kormendi, R. C., and Meguire, P. 1985. "Macroeconomic Determinants of Growth: Cross Country Evidence." *Journal of Monetary Economics* 16:141–63.

Leonard, H. J. 1988. *Pollution and the Struggle for World Product*. New York: Cambridge University Press.

McGuire, M. 1982. "Regulation, Factor Rewards, and International Trade." *Journal of Public Economics* 17:335–54.

Nordhaus, W. D. 1989. "The Economics of the Green House Effect." Yale University. Mimeograph.

Pethic, R. 1976. "Pollution, Welfare, and Environmental Policy in the Theory of Comparative Advantage." *Journal of Environmental Economics and Management* 2:160–69.

Ridgeway, J. 1970. *The Politics of Pollution*. New York: Dutton.

Tobey, J. A. 1990. "The Effects of Domestic Environmental Policies on Patterns of World Trade." *Kyklos* 43:191–208.

Walter, I., and Ugelow, J. 1979. "Environmental Policies in Developing Countries." *Ambio* 8:102–9.

World Resources Institute. 1989. *World Resources 1989–90*. New York: Oxford University Press.

World Resources Institute. 1990. *World Resources 1990–91*, New York: Oxford University Press.